U.S. 40 TODAY

Thirty Years of Landscape Change in America

THOMAS R. VALE and GERALDINE R. VALE

THE UNIVERSITY OF WISCONSIN PRESS

Published 1983

The University of Wisconsin Press
114 North Murray Street
Madison, Wisconsin 53715

The University of Wisconsin Press, Ltd.
1 Gower Street
London WC1E 6HA, England

First printing

Printed in the United States of America

For LC CIP information see the colophon

ISBN 0-299-09480-4 cloth
ISBN 0-299-09484-7 paper

FOR

George R. Stewart

(1895–1980)

Contents

Preface

IN 1953, Professor of English George R. Stewart published *U.S. 40,* in which he presented ninety-two scenes that he had photographed along the route of U.S. Highway 40 from Atlantic City to San Francisco as a "cross section of the United States of America." He selected views which he felt represented typical landscapes along the route, rather than ones that were necessarily scenic or otherwise attractive. For each photograph Stewart wrote a short interpretive essay in which he discussed why the landscapes had the features that they did. Stewart concerned himself with both natural and cultural characteristics, and often stressed the genesis of those characteristics, including human history wherever he felt that it helped mold the personality of the place. His book remains a fine example of landscape portrayal and interpretation.

What held the photos together, and gave the book its unity, was the highway. But although Stewart was obviously intrigued by asphalt pavements, steel bridges, and street signs, he was concerned with much more than the mere physical structure of the road:

> We must consider all that it means to the man who drives along it. It must be not only what can be seen, but also what can be felt and heard and smelled. We must concern ourselves with the land that lies beside it and the clouds that float above it and the streams that flow beneath its bridges . . . the people who pass along it and those others who passed that way in the former years. . . . Only by considering it all, as we drive from the east or from the west, shall we come to know in cross section, the United States of America.

Stewart, then, saw his book as exploring two separate subjects, the highway and the route it followed.

In Stewart's day U.S. 40 was a major transcontinental highway bisecting the country. A traveler could follow its familiar black-on-white shield for 3,000 miles from coast to coast. Today the longest stretch for which one can follow the U.S. 40 emblem uncompromised by any competing interstate insignias is for the 440 miles between Denver and Salt Lake City. In other places it shares recognition with the newer designations of Interstate 70 or 80. In still other places it has become a secondary regional or local highway or even a frontage road. In fact, for its entire length in Nevada and California, U.S. 40 has yielded its identity as a marked highway to Intersate 80; there its shield hangs only in the few spots where it has escaped the notice of highway crews, as it apparently has on at least one side street in Reno, where it formerly directed lost tourists back onto the route west. But whether U.S. 40 exists officially today as a transcontinental highway or not, its route can still be followed. Moreover, the viewpoints from which Stewart snapped his shutter are not only traceable today, but may be relocatable for many years to come.

The idea of tracking down the sites of Stewart's photographs has occurred, we are sure, to many persons who have enjoyed *U.S. 40*. But, as far as we know (and Mr. Stewart confirmed our suspicion), by 1979 no one had systematically retraced his tire tracks and footsteps across the continent. We decided to meet that challenge, to rephotograph the scenes, and to use the resulting photo pairs as the basis of a study in the changing landscapes of America. The idea that a photograph captures only one static view of a constantly changing landscape is a notion that was also apparent to Stewart: "In describing some of the pictures, I have mentioned the imminence of change, but a continual repetition would become monotonous, and on the whole it seems better merely to write a general warning: 'Thus it was when I passed by, in my time.'" Observing and writing about the evidence, rather than merely the imminence, of change, we experienced an excitement quite remote from monotony. This book, *U.S. 40 Today,* enables the reader to see not only how it was when Stewart passed by, but also how it was, a generation later, when we passed by.

In the pages that follow, we have included seventy-two of Stewart's scenes, the others omitted for a variety of reasons. For example, several of Stewart's photos feature people rather than landscapes, and thus do not suit our purposes for *U.S. 40 Today*. In addition, some views seem repetitive and offer little that is not found in the photos we have used. We were also hesitant to climb over a KEEP OFF sign and up a precarious ladder on a tall water tower. Finally, the negative for one scene is apparently unavailable. For each view we have chosen, Stewart's photograph appears at the top of the page and ours at the bottom. Accompanying the pairs of pictures are short essays which discuss the changes that have occurred in the scenes and their environs since Stewart's time. We have tried to evaluate changes both conspicuous and subtle, both natural and cultural, both "attractive" and "unattractive." The photographic scenes represent changes of differing scales, some ephemeral or local, others persistent or widespread. They illustrate not only changes arising from conscious human attempts to alter conditions, but also changes that are the unanticipated results of human activities. Some of the scenes reveal only minor changes, and a few show essentially no change. All encourage reflection on attitudes toward landscape change.

How well does *U.S. 40* provide the basis for a study of landscape change? As a group, Stewart's photographs portray a variety of subjects that make the photo comparisons more informative, and more interesting, than if they were predominantly scenic vistas or historic buildings. Nonetheless, we feel that Stewart treated a few topics unevenly. For example, in parts of the East he seemed most concerned with human history, occasionally using pictures of historic places that opened a door to the past but prompted little or no commentary on the present. For these pictures, Stewart simply related stories of historic events or persons. In Nevada, on the other hand, Stewart seemed preoccupied with the vast vistas of the Nevada landscape. He ignored other interesting subjects, such as highway crossings of the Humboldt River, and offered no intimate looks at small towns. Given the choice, we might have included, for example, U.S. 40 crossing the Missouri River in central Missouri, the first view of the Rocky Mountains from the High Plains, a street scene in the railroad town of Battle Mountain, Nevada, or the orchard district near Fairfield, California.

On the other hand, in general, Stewart did not underrepresent the small American town, as some critics have suggested, or the large city, as Stewart himself alleged. We randomly located ninety-two points along the route of U.S. 40 of 1950, and landed in towns and cities only eleven times. Stewart had twice as many scenes, twenty-two, in such settings. Stewart's disproportionate attention to towns and cities is not inappropriate, given their importance to people and human history. His generally good choice of subjects in terms of their geographic and demographic importance helped to produce a book that is a fine basis for a study of landscape change.

We have adopted some conventions. Stewart took most of his photographs in two coast-to-coast highway trips in 1949 and 1950, although the date of any particular photo cannot be easily determined; for simplicity's sake, we refer to all of his photos as taken in 1950. Similarly, we describe our own pictures as representing 1980, even though many in the western states were taken in 1979 and a few in 1978. We made great efforts to find as exactly as possible the very spots where Stewart had stood with his camera, and we were surprisingly successful. However, if a precise duplication of his view had become obstructed by vegetation or obliterated by dynamite, or if it omitted something of interest, we felt free to move a few steps or use a lens with a wider field of vision; in these cases we refer in the text to the differences. Moreover, cropping and printing requirements sometimes prevented the presentation of precise duplication of Stewart's views. Furthermore, our purpose is to assess landscape change over the last thirty years, and not to duplicate Stewart's objective of interpreting the characteristics and origins of features more generally. Thus, many things which Stewart discussed we do not, and vice versa. Our book is different, and stands alone. Yet, we are particularly interested in comparing Stewart's impressions of places with our own, and in evaluating his predictions about what was for him the future. Therefore, we quote freely from *U.S. 40,* and quotations otherwise unidentified are taken from that book. We also retain the titles of the photos which Stewart used. Moreover, as Stewart often included personal reflections on what he experienced, we too comment about the happenings along our way. This book, then, is not simply an objective record of physical change. It is also a journal that chronicles the experiences of two travelers as they come to know better the America of 1980 through the perspective made possible by a most astute sojourner of 1950, and the stories that the photographs unfold are varied, rich, and full of land and life.

We thank the American Philosophical Society, which aided the 1980 field work with a grant from the Penrose Fund. Mrs. George R. Stewart kindly granted permission for our extensive quotation from *U.S. 40.* Prints of most of the Stewart photographs were provided through the courtesy of the Bancroft Library of the University of California, Berkeley. We also acknowledge the support of geographers Peirce Lewis, Donald Meinig, James Parsons, James Shortridge, and Ian Simmons, and of Peter Givler, Elizabeth Steinberg, and Brian Keeling, all at the University of Wisconsin Press. The University of Wisconsin Cartographic Lab prepared our graphic materials.

U.S. 40 TODAY

Roads and Motoring in America Today

CHANGE HAS BEEN characteristic of roads and road travel in America. When Stewart wrote *U.S. 40,* he commented about the changes which had prompted him to follow U.S. Route 40 rather than one of the named highways well known earlier in the century:

> Some people ask: "Why did you not choose the Lincoln Highway?" The answer here is easy: "The Lincoln Highway is dead!" The name, to be sure, survives in local usage. But the Lincoln Highway Association closed its offices in 1927, and most of its red-white-blue rectangles have long since been removed from the highways to make way for the white shields of the current national system.

Today's traveler, probably unfamiliar even in "local usage" with the Lincoln Highway, might feel that it is the numbered U.S. routes, like U.S. 40, which are "dead" as the major highways crossing the country. Their place has been taken by the modern interstate freeways, which not only carry the bulk of long-distance vehicle traffic, but also dominate or at least represent the character of the contemporary experience of driving. Their familiar multicolored shields have replaced the black and white insignias of the U.S. route system on many major highways and in most American minds.

ROADS

The routes followed by most roads in America reflect conscious human choice. Those choices, however, involve two different histories.

Some roads evolved from footpaths that were gradually widened into horse trails and then into cart tracks. The segments of U.S. 40 in New Jersey and eastern Maryland probably had such an origin. Similarly, the 500 miles of U.S. 40 between Wells, Nevada, and Sacramento, California, began as the California Trail, by which mid–nineteenth-century travelers moved to the mines and farms in the Golden State.

Other roads, although perhaps following the general routes of a few early explorers or emigrant parties, were built specifically for vehicles and had neither paths nor crude wagon trails as evolutionary precursors. The stretch of U.S. 40 between Salt Lake City and Wells largely represents a route avoided earlier because it lacked water; it was constructed as a modern highway only in this century. Lengths of U.S. 40 across eastern Colorado and much of Kansas began as roads built parallel to the already established rail lines.

The earliest, and in many ways most important, segment of U.S. 40 originally built as a road is that between Cumberland, Maryland, and Vandalia, Illinois. This stretch of highway is the route of the historically famous National Road. It was financed by the federal government in the first decades of the nineteenth century and served to link the Ohio and upper Mississippi Valleys with the East Coast. Even though the National Road represents only a fourth of the total length of U.S. 40, many people think of the two as synonymous. This assumption reflects the importance of the National Road in the expansion and solidification of the early United States. In part, however, the view that the National Road was the precursor for U.S. 40 is an eastern bias. In the West, U.S. 40 follows the California Trail for a distance about as long as the National Road. Moreover, the existence of the California Trail facilitated the linking of California to the nation in much the same way as the National Road had earlier tied the Midwest to the eastern seaboard. Californians may well view the precursor of U.S. 40 as the California Trail rather than the old National Road.

The various segments of the route of U.S. 40, then, had different origins. They became joined together as a single, numbered route as a result of the rise in importance of the private automobile.

Early in this century, the emergence of automobile travel increased the need for usable roads. Muddy quagmires, too few directional signs, unsafe bridges, inadequate regulation of urban traffic—all were hazards of early travel as wagon trails became used by increasing numbers of motorized vehicles. Primary responsibility for improvement of roads rested initially with local governments, which understandably concentrated their efforts on surfacing urban streets or otherwise providing for regional traffic. Little consideration was given to roads for long-distance travel.

Business people provided the earliest impetus to the development of interregional and even transcontinental routes with their formation of private highway associations in the second decade of this century. Financed by money donated by owners of restaurants, hotels, and other services which catered to the traveler, each highway association promoted the improvement of its individual route, hoping thereby to capture a higher proportion of the traveler's dollar than some competing road. These routes were named with such historically or geographically evocative titles as "The Old Spanish Trail" or "The Dixie Highway," and each was identified by a particular sequence of colors, such as red-white-blue or red-white-yellow, painted in bands on utility poles. Stewart explained well the situation as it developed in the early part of this century:

> After a few years—by 1920, let us say—the situation was growing intolerable and ridiculous. In the first place, the contributions levied upon the local business-men went chiefly into publicity and salaries of the so-called "highway officials," not into good roads, or even to any great extent into good marking for roads. . . . Besides, the "trails"—as they were commonly called—were getting too numerous. Sometimes the poles were painted almost from top to bottom. On certain roads as many as eleven different trails coincided. To spot the proper colors under such circumstances was something of a problem.

This chaos, and the question of whether public money for highway improvement was being spent in ways best serving the auto traveler, were evaluated by a Joint Board of Interstate Highways, involving state and federal officials, in the 1920s. In 1925, the Board established the system of U.S. numbered highways which followed major roads across the states, although not necessarily adhering to the named "trails." East–west routes were given even numbers increasing in value from U.S. 2 along the Canadian border southward to U.S. 90 along the Gulf Coast. North–south routes were given odd numbers increasing from U.S. 1 along the east coast to U.S. 101 along the west coast. These highways were not designed, constructed, or financed by the federal government. Rather, the U.S. routes remained primarily the financial and administrative responsibilities of the states, although federal aid has existed since at least 1916. U.S. 40, then, was one of the "new" numbered U.S. routes, following parts of the National Old Trail, the Victory Highway, and sections of other named roads.

Three years after Stewart's book was published, the autonomy of the U.S. numbered routes, including U.S. 40, came into jeopardy when the interstate system was authorized by the 1956 Federal Aid Highway Act. In fact, today fifty-two of the photo views in *U.S. 40* are on or along routes appearing on road maps as Interstate 70 or 80. The contemporary interstate system in some ways is simply an elaboration upon the characteristics of the older numbered U.S. highways, but in other ways is quite different. The money for the interstates was generated from user taxes and fees, as had been the practice for roads generally on both the state and federal level. Moreover, as with the earlier federal highway aid acts, the 1956 law provided that the money be given to the states, which were then responsible for the awarding of the contracts for actual construction. The 1956 Federal Aid Highway Act thus continued the federal role as one of coordination and overview, as well as of financial assistance.

The numbering of the interstates was also similar to that of the U.S. routes, except that the progression from small to large numbers across the country was reversed, presumably to reduce possible confusion between the two systems. Thus, the east–west interstates carry even numbers, increasing northward, with the major routes designated 10, 20, 30, and so on. North–south interstates have odd numbers, increasing eastward, with the main routes ending in 5.

Unlike the earlier federal aid highway acts, however, the law authorizing the interstates involved a stronger federal role. Federal funds have provided about 90 percent of the construction costs for the interstate highways, and only 10 percent has been provided directly by the states; thus the interstates are more truly a federally financed system than the U.S. numbered routes. Also, the interstate system was the first attempt to design and construct new highways to cross the country rather than merely to designate routes along existing roadways. The limits on access, the multiple lanes, and the freedom from tolls all were required features of the interstate system, characteristics designed to facilitate long-distance travel.

Being new highways, the interstate freeways are less bound to the routes of the past than were the highways of the U.S. route system, which were, in Stewart's mind, too constrained by inertia:

> The pavement may have progressed from dirt to macadam to oil to thin concrete to thick concrete, and the route may have been straightened and made less steep, and the right-of-way may have been gradually widened. At length and after lamentable wastage of money, the path may have become a freeway, and yet the freeway may still bend here and there a little, because the path so bent. . . . The numbered transcontinental highways already begin to show something historical and quaint in their routes, looking back toward the time when there were few paved roads, and when even the most forward-looking planners thought in terms of only a few hundred miles between cities, not really of a transcontinental road.

Still, not even on the interstates is the traffic primarily "transcontinental," and a highway designed only for vehicles traveling between, say, New York City and San Francisco would be little used. Interstates, like any major through-highways, must also serve the needs for regional and even local movement. Route selection for major roadways like the interstates is always one of compromise between alignments and designs which favor the truly long-distance travelers, who are relatively few, and those that serve the regional travelers, who are much more numerous. The interstates, then, are not truly free from the past, for they too angle toward existing towns, shy away from rough terrain, and follow alignments of existing roads. They too "bend here and there a little, because the [earlier paths] so bent."

By 1980, 95 percent of the 41,000 miles of interstate freeways had been completed. The total expenditure of eighty billion dollars was, however, far greater than the initial estimated cost of twenty-seven billion. In addition, the currently projected cost of the final 2,300 miles of the system is thirty billion dollars, with funding from user taxes declining because of the greater fuel efficiency of modern vehicles, lessened auto traffic prompted by high fuel costs, and general stagnation of the economy. Moreover, some sections of the interstates already completed are beginning to show the potholes and crumbling surfaces produced by heavy use, particularly by trucks. Simple maintenance costs now loom as an unforeseen financial burden for some states.

Stewart anticipated much of the physical change in the design of the highways following the route of U.S. 40. Indeed, the trends were already apparent in 1950: "If the present indicates the immediate future, we may say that the development [of U.S. 40] is progressing along three lines. It is becoming a four-lane highway; it is being transformed into a freeway; it is bypassing cities and towns." Today, 2,000 miles of the 3,000-mile route have at least four lanes, 1,500 miles are freeway, and all major cities are bypassed. Stewart even anticipated the possible change in the designation of the route of U.S. 40, just as he did in the design of the highway itself: "As long as our civilization develops and amplifies, U.S. 40—by whatever name it may be called—will also develop and amplify." The route of U.S. 40 today is followed by Interstate 70 between Wheeling and Denver, and by Interstate 80 between Salt Lake City and San Francisco. The absence of a single interstate highway coinciding with U.S. 40 from coast to coast reflects the differences in the origins of the two federal highway systems and in the patterns of population and traffic which each system was intended to serve.

MOTORING

The development and amplification of motoring roads, as typified by the construction of the interstates, have altered the experience of driving over the last thirty years. Three themes come to mind.

First, the interstate freeways, like all freeways, seem removed or even isolated from the landscapes through which they pass. The wide right-of-way, the absence of roadside businesses, the massive cuts and fills, and the ever-present chain-link, barbed wire, or woven fence all contribute to the lack of intimacy between road and setting. The land beside a freeway seems far away, cut off from the pavement hurrying travelers to distant destinations. A more modest highway, like the U.S. 40 of Stewart's day, is close to the landscape features beside the pavement, and permits the traveler to appreciate better "the land that lies beside it . . . and the streams that flow beneath its bridges." Roadside settlements, small towns, and large cities alike assume a more vivid reality if one is forced to slow down and pass beside or through them rather than to rush on around them. We welcomed, for example, the slower pace and change of scene offered by the movement through the glittering towns of Nevada which the freeway did not yet bypass in 1980. On the other hand, we occasionally took advantage of the freeways to bypass the most congested streets of such cities as Kansas City and St. Louis.

The increased isolation typical of the interstates is similar to the character of what Stewart called a "dominating highway": "A dominating highway is one from which, as you drive along it, you are more conscious of the highway than of the country through which you are passing. Six-lane highways, and four-lane highways, particularly in flat country, give this impression." Yet, Stewart praised the portion of U.S. 40 that was freeway in 1950: "The freeway now extending nearly all the way between Sacramento and San Francisco, seems to represent close to the ultimate in motoring. It possesses the advantages of a parkway, and yet passes through farmlands and natural countryside that have not been artificially prettified."

Perhaps in 1950 too little of the total length of U.S. 40 was freeway for Stewart to consider such a road design to be intrinsically dominating; or, on the other hand, because the perception of a highway as dominating is tied to the road's setting, Stewart may have felt that in rugged high mountains, or even gently rolling hills, a freeway could not become so dominant as to be the focus of attention. Still, we think that a freeway is more likely than a smaller road to be removed from its setting, and is thereby less likely to be what Stewart called an "equal" highway, "one which seems to be an intimate and integral part of the countryside through which it is passing." We admire, for example, the eastern ascent of two-lane U.S. 40 over Berthoud Pass in Colorado, and feel that its conversion to a freeway would eliminate the feeling of closeness to the mountain slope and detract from the exhilaration of climbing up the switchbacking route over the high ridge. Similarly, the four lanes of non-freeway U.S. 40 across Indiana provide a more intimate impression of the green fields of corn and soybeans and the neat white farm houses than the four freeway lanes of the parallel Interstate 70.

A second characteristic of modern motoring, which contributes to the feeling of isolation, is the high speed of travel. Rushing on at fifty, sixty, or seventy miles an hour, the traveler cannot observe the detail of features beside the pavement or even close to the right-of-way. Thus, the driver's attention is focused either on the road ahead or on the distant view, with only fleeting and blurred glimpses of weeds on the shoulder, birds on the right-of-way fence, farmers in nearby fields, or buildings on side roads. The geographer Carl Sauer lamented the speed of modern travel, and urged his students to pass slowly over landscapes: "Locomotion should be slow, the slower the better, and should be often interrupted by leisurely halts to sit on vantage points and stop at question marks."[1] We appreciated all the halts occasioned by the "question marks" of Stewart's "vantage points," even when those stops were in landscapes often considered common or unsightly. A vantage point that particularly prompted our thanks for slow locomotion was the seat on the bridge at Donner Pass. What modern freeway would feature such a concession—or seduction—to leisure?

What is the best speed at which to travel? There is, of course, no single best speed, for each offers something a little different. Walking slowly allows the inspection of an individual old oak, or an edge of a wheat field, or a weedy and littered road shoulder, or the architecture of a Main Street. "Poking along" at twenty or thirty miles an hour presents opportunities to compare changes in vegetation on different slopes, investigate the crops growing in adjacent fields, scare up birds from the right-of-way, or examine the activities beside side streets of a small town.

Freeway speeds may not allow these slower-paced pleasures, but they do have pleasures of their own. Quite apart from offering the obvious advantages of getting from point to point in little time, the rapid progression over the land facilitates comparison of different landscape features. Along a freeway, dissected hills (hills cut into by streams) with woods and pastures may give way to the urban landscape of homes, stores, and factories, only to yield to vast plains of row crops and farms, all within a matter of minutes. The landscape becomes, in such travel, a kaleidoscope of images. Moreover, the perspective on a given feature changes rapidly, often encouraging an appreciation of the importance to the perception of landscape elements of point-of-view, of light and shadow, and of proximity. A pointed lone mountain peak, as seen from one side, may become a long squat ridge when seen from the other. A grain elevator may stand out sharply one minute when a traveler views it with strong back-lighting, but the same elevator may nearly disappear the next minute when seen with flat forelighting. As a distant ridge is approached from afar, it may appear at first as an apparition on the horizon, then as a dark somber slope, then as a forest of green pines, until finally, as the road ascends the ridge, the trees may be seen as growing amid a grassy and flowery understory.

Each point on the continuum of speeds and perspectives from leisurely strolling along a roadway to hurtling through Earth's orbit in a spaceship has its own rewards. Travelers on a 600-mile-an-hour jetliner over the country's midsection, for example, may ponder the checkerboard pattern of crops 30,000 feet below and come to appreciate the grid survey system by which the landscape is divided. They may also wonder about the circular fields that appear from time to time on the High Plains and, if the day is especially clear and the season summer, see that these circles are expanses of crops irrigated by rotating, central-pivot sprinklers. Many landscape-watchers shun freeways, but perhaps they need to cultivate the sensitivity appropriate for such motoring, rather than to assume that no amount of sensitivity would help. We think that Stewart would have agreed.

A third characteristic of the modern freeway, one that makes possible uninterrupted high speeds over long distances, is the general uniformity in road design. One can drive from coast to coast on a multilane highway without a single intersection, stop sign, traffic signal, or railroad crossing. If one could arrange for refueling while traveling, a nonstop trip from coast to coast would be possible. In Stewart's day, by contrast, cross-country travel involved freeways, four-lane divided highways, four-lane nondivided highways, three-lane highways, and two-lane roads. But the standardization in roadways today may be more apparent than real. Indeed, the interstate freeways do vary from place to place, reflecting not only differences in design preferences among state highway departments but also design changes from one period of construction to another. The variety is in such features as median form, guardrail design, bridge width, and sign placement. A contemporary traveler could easily develop the skills to admire a cloverleaf overpass, as Stewart did an S-bridge in Ohio; a modern freeway roadside rest, as Stewart did a primitive roadside park in Kansas; the variable width in the freeway shoulder, as Stewart did on the two-lane road in Nevada; and a right-of-way fence, as Stewart did nearly everywhere.

Apparent uniformity is not restricted to the modern roadway itself; it also characterizes the adjacent contemporary service for food and lodging. Whereas

Stewart had the Red Brick Tavern in Ohio and a small auto-court near Emigrant Pass in Nevada, we had McDonald's and Motel 6 in innumerable cities and towns. The considerable advantage of uniformity in eating and sleeping establishments is the security that comes from knowing beforehand the character of goods and services. Everyone is assured of a predictable Big Mac regardless of the particular golden arches under which it is prepared, and in a Motel 6 everyone can be confident not only of the price but also of the floor plan and even the color and design of the bedspread. The disadvantage of uniformity in roadside services is a loss in the richness of experience, whether involving local or regional meal preferences, personal attention or neglect in lodging, or the intimate interaction with people that so often is associated with family-run businesses. The trade-off is familiar because it is often characteristic of "progress": on the one hand, we gain the convenience and comforts that attach to predictability; on the other hand, we lose the sense of adventure and the stimulation of human contact.

The contemporary cross-country traveler who spends much time on the interstate freeways, then, remains more isolated from the landscape, progresses more rapidly, and experiences greater uniformity in food and lodging than the traveler of thirty years ago. One may choose, of course, to drive on the back roads or at least on the major highways which are not freeways, and stop at family diners or quaint lodges, in order to capture the feeling of an earlier era. Many people do. But the interstates are still there, and it is often hard to resist their attractions. Some of our friends have suggested that Stewart's day was a kind of golden age of motoring because roads were good and abundant, travelers were few, and costs were low. Perhaps, however, people today live in a better world than did Stewart, for motoring at least, because they can select the type of traveling that suits their desires and purposes, which may change from time to time or place to place. Opportunities for choice are certainly greater.

Other characteristics of motoring have not changed. Regardless of the nature of the highway or of the associated experiences of auto travel, the pleasures of driving remain much as Stewart described them:

> There are the visual sensations. . . . The low sun ahead with the shadows pointing this way, and the glare in the eyes. The sun high above, and the light full on the road with the shimmer of heat-waves rising from the asphalt. . . .
>
> There are the sounds. . . . The whir of the tires on concrete. . . . The soft squish of wet tires on pavement running with rain. The continual whirl and hiss of air around windows. . . .
>
> There are the smells. . . . The freshness of early morning before the day has matured, and that other freshness that comes after the thunderstorm. Now and then, the gasoline odor, or the diesel odor, as a truck goes by. And the smell of the car itself too—as comforting as home . . .
>
> There are, too, those vague sensations that we call kinesthetic. . . . All the rivers of fresh air coursing over the face. The pressure backward with acceleration, and the pressure as the body swings forward when the brake goes on. And the continual joggling from the springs, doubtless good for the digestion and the nerves . . .

These sensory experiences peculiar to motoring cannot be had except by climbing in behind the wheel. What we offer in this book, in pictures and words, is a look at the changing landscapes of America, through space and through time. This presentation will start, where Stewart started, on the East Coast where the waves of the Atlantic Ocean lap against the famous boardwalk of Atlantic City.

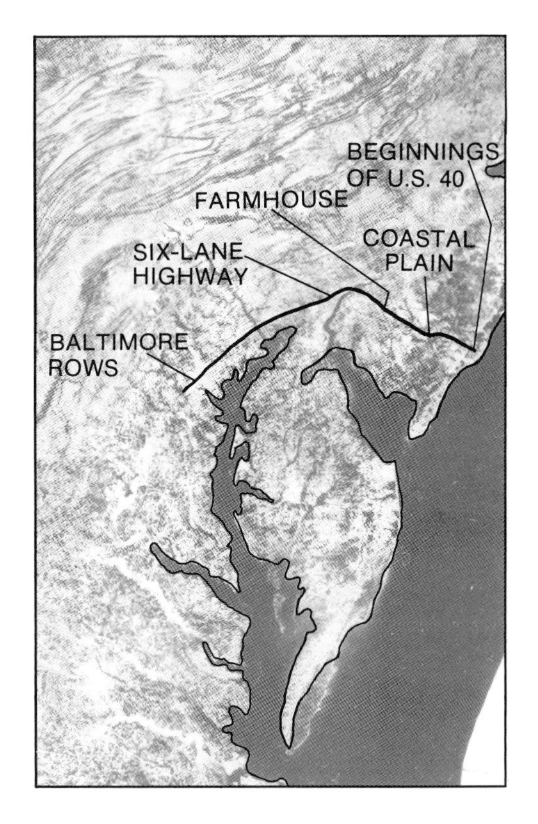

ATLANTIC CITY TO BALTIMORE: ATLANTIC COAST

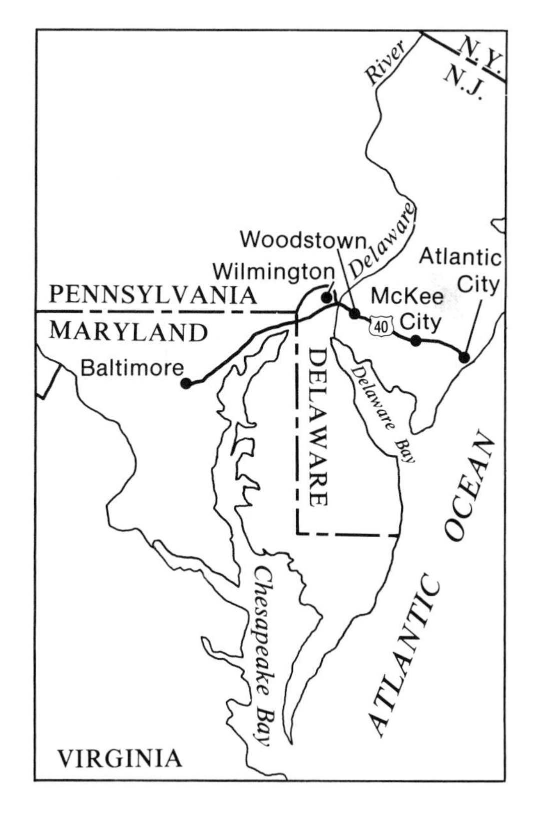

Atlantic City to Baltimore: Atlantic Coast

THE FIRST 140 MILES of U.S. 40 from Atlantic City, New Jersey, to Baltimore, Maryland, are best envisioned in two segments, and the two are as sharply different today as they were in Stewart's day.

Westward from the Atlantic Ocean, the highway crosses the nearly flat terrain of the coastal plain. It still "follows a causeway across the salt-marshes of the coast, goes through the scrub-pine belt, rises a few feet into a country of fields and pastures and woodlots, dips down again to tide-level." Most of the landscape remains rural, either in forest or in agricultural land of corn and dairy cattle. West from McKee City, the highway is only two lanes, with most vehicles using the multilane highways of the Atlantic City Expressway and U.S. 322. Stewart's assessment that "you will find few pleasanter little runs that the 68 miles westward from Atlantic City" seems just as valid today. The highway is "a delightful little two-lane road. . . . not heavily traveled, having few straightaways for such flat country, and curving intimately among almost imaginary hills and through quiet towns and villages."

In contrast to this initial rural stretch, U.S. 40 from near its crossing of the Delaware River all the way into Baltimore is an urban and suburban highway. Stewart's comments from 1950 again capture the essence of this second segment today: "The highway itself dominates. . . . With the scenery unimpressive, there is little to watch except the four lanes with their constantly changing traffic, and the roadside succession of service-stations, 'joints,' and tourist courts." The river itself is spanned by a bridge that was just being completed when Stewart published his book, thereby eliminating the ferry which formerly linked New Jersey and Delaware. Southwestward from the bridge, U.S. 40 now is the business route into Baltimore, being paralleled by the major interstate freeway along the Atlantic Coast, I-95, which carries most of the through traffic.

This demotion of the importance of U.S. 40 by the interstate highway along the northern edge of Chesapeake Bay is the first example of a relationship that repeats itself across most of the country. In Stewart's day, U.S. 40 was a major east–west thoroughfare, a road which not only carried most families on vacation trips and truckers on business hauls but also created the economic base for the towns and service establishments which it linked together. The location of old U.S. 40 remains a vital roadway, but the interstates which today follow the route carry the bulk of the vehicles. New businesses have sprung up in response to the modern patterns of travel, while the old cafes, gas stations, and motels may suffer or even die. U.S. 40 as a designated highway, is often the city street, the local route, or even the frontage road. Its glory belongs to the past.

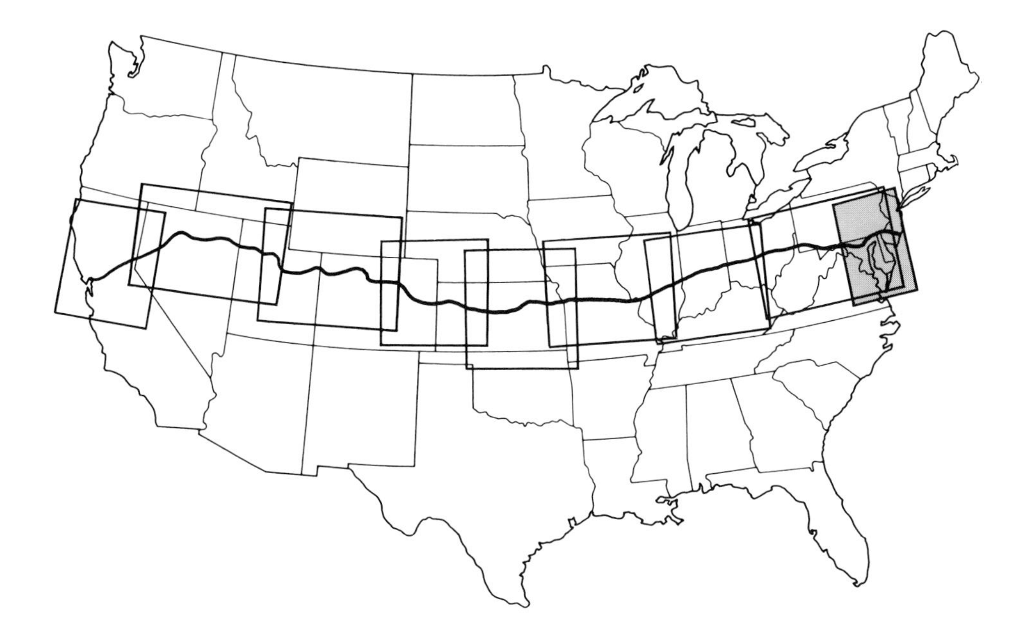

BEGINNINGS OF U.S. 40

Beginnings of U.S. 40

The view is westward, across Atlantic City, from a point within several hundred feet of the Atlantic Ocean. At first glance, the two photos, although obviously taken from different spots, reveal so few changes that someone unacquainted with the facts might think that they were taken on the same day. Imagine the later photograph to have been taken first, on a clear morning; the decreased light intensity of the upper photo could be explained as the result of the afternoon clouds having thickened and expanded, obliterating the sun and dulling the appearance of the buildings and of the War Memorial in the traffic circle. The clock on the tower of the high school indicates that time separates the two photographs, but the interval could be as short as ninety minutes, rather than thirty years. The sturdy frame houses appear neat and well maintained in each picture. Neither the residential streets nor the commercial strip indicates much alteration by new construction. Beyond the Intracoastal Waterway, the athletic stadium and airport retain their modest, small-town appearance, and farther west the emptiness of the coastal marshes extends to the horizon.

The lack of conspicuous modification of the landscape in this view, however, belies recent changes elsewhere in Atlantic City that have occurred over the last thirty years. The necessity of taking the recent photo from a location different from that of Stewart's suggests the scope of this transformation. Stewart stood on the roof of the Roosevelt Hotel, located at the end of the major highway coming from the mainland. In 1979 the structure on this prime site was spectacularly destroyed with explosives, and the lot was subsequently cleared. A multistoried gambling casino will eventually rise and join the other such businesses which now are scattered along the famous boardwalk. The building from which the 1980 photo was taken, the Mayfair Building, now houses not only hotel tenants but also the offices of still another casino which is already being constructed immediately to the east. We shared the elevator with one Mayfair resident while we rode up to our picture-taking site. She summed up her reaction to the demolition and construction which is driving up the costs of living and thus creating hardship for the large numbers of retired folk: "They'll be tearing this one down next."

All along Pacific Avenue, at right angles to the picture view and extending in front of the Mayfair Building, construction is common. Atlantic City, born in 1854 as a seaside resort town for urban dwellers farther north, is now experiencing renewed commercial activity as a result of the legalization of gambling by state voters in 1976. Restricted to certain few locations, the gambling rooms, though crowded with patrons, seem quiet, even hushed, as if the people were a little self-conscious of their activities. The reserved atmosphere of the casinos of Atlantic City, New Jersey, is far removed not only spatially but culturally from the noisy bustle of the casinos on U.S. 40 in Reno, Nevada.

In the photo view some subtle changes are evident. The trees in the small park to the right have matured, although they remain only sparsely foliated. The neatly trimmed rectangular shrubs are fewer, and benches have been placed beside the street in front of the school where newly planted trees will eventually provide shelter. Lane markings and islands are conspicuous in the recent photo, and channel the greater traffic more rigidly than in Stewart's time.

As in 1950, U.S. 40 simply "emerges from the city streets" and lacks a definite terminus. The first route sign appears at the end of the straightaway, and from that point the familiar shield can still be followed westward out across the continent almost as far as Salt Lake City.

COASTAL PLAIN

LOOKING EAST from the McKee City fire tower, in the recent photo, we see U.S. 40 on a Sunday evening in midsummer. Most of the traffic is westbound, presumably weekend vacationers heading home to Newark, Philadelphia, Baltimore, and innumerable other cities and towns farther inland. Long shadows from poles, trees, and the fire tower itself suggest the lateness of the hour. The tall buildings of Atlantic City, visible in Stewart's photo, are lost in the low light and evening haze in our picture.

The traffic is heavy, but most of the landscape through which the busy highway passes is sparsely settled forest. Stewart astutely anticipated that the "thin soil" of this oak and pine woodland would likely preclude agricultural development, and thus the region would "remain permanently forested." The continued undeveloped condition and state ownership of large areas of these pine barrens have made them, in fact, a region admired and used by wild landscape recreationists. Even Congress has recognized the importance of this open space so close to urban centers by authorizing a one-million-acre reserve in similar country north of the photo site.

The land away from the roadway may be empty, but that immediately fronting the highway is not. Stewart regretted the newly established motel in the older photograph as the precursor of a strip of tourist facilities which would impede "the free flow of traffic." Today, motels, restaurants, and other tourist facilities line the highway. Billboards, set back in areas cleared of trees and shrubs, occupy the land between the businesses. The power line providing electricity for roadside development and for the street lighting necessitated by the increased congestion adds to the cluttered look along the highway.

McKee City is not, however, a booming tourist town. Established and named by land developers early in this century, McKee City is little more than a scattered strip development along the highway, a traffic circle, and two racetracks. The tourist businesses cater to the patrons of horse racing so exclusively that, in the words of a commentator on the New Jersey scene, they "know the pleasant jingle of cash only when the ponies run."[2] We found the largest restaurant complex along this stretch of highway to be closed, in fact, on a Monday afternoon in July, but we assumed that when the horses are running both the large parking lots and the valet service are much needed. McKee City could eventually become the urban center dreamed of by its creators, but it would require that present urban dwellers would want to move, and that fast, cheap transportation for commuting would be available. For now at least, the landscape remains largely wild forest and bog.

In spite of the lure of Atlantic City, the highway itself is virtually unchanged. Pavement materials are today, as they were before, concrete for the westbound lanes and asphalt for the eastbound lanes. The road alignment and design are unchanged. The need for a freeway, which Stewart foresaw in 1950, was satisfied not by rebuilding U.S. 40 but by constructing a parallel route one-half mile to the north of the existing roadway; it is hidden from our view by the dense vegetation. U.S. 40 has become, as a result, a secondary road rather than the major highway through the pinelands that it was in Stewart's time. In varying degrees, this decrease in the status of U.S. 40 over the last thirty years is evident along much of its route.

FARMHOUSE

Farmhouse

West of the pine barrens, the sandy soils give way to finer-textured loams, and the empty forests yield to fields of alfalfa and corn, of tomatoes and asparagus. The stretch of highway in this view is crossing the gently rolling country just east of Woodstown, New Jersey, a landscape which reminded us of the upper Midwest: crops of field corn and hay, pastures with milk cows, and farms with silos.

The farmhouse, which Stewart described simply as "a good early American house," is home for a dairy farmer who pastures his herd to the right of the picture and who may harvest the alfalfa field across the highway from his home. The maple tree in the front yard has grown so large that our view of the house is blocked. The willow in the backyard is gone, but a small tree has grown up on the fence line east of the garden plot of corn. Additional tree growth on the horizon near the highway gives testimony to the ease with which woody plants grow in this moist mild climate and these fertile soils. The hedge fronting the house is as perfectly trimmed in 1980 as in 1950, although it has grown sufficiently to provide privacy for the farming family and perhaps to block traffic noise. The pointed and conical "top knots" adorn the hedge as more recent curious embellishments. The barn has been painted white since Stewart's time, but its generally poor condition is not revealed by the photograph. With broken windows, sagging doors, and peeling paint, the barn is now being used to store field machinery and hay. The farm's small silo is behind the barn, hidden from our view, although other outbuildings are visible in the backyard; these probably provide interior space for milking and shelter for the cows. The house itself is in good repair, but the window shutters and front porch, both noted by Stewart, are gone. The mowed grass on the slope between the hedge and the highway suggests that the family takes pride in its modest farmstead.

Reflecting a kind of permanence appropriate for this farming landscape, the utility lines occupy the same alignments they did in 1950. The electricity line continues on the right side and the telephone line, now with cables instead of individual wires, is on the left. The poles are new, however, and seem to be set in different locations.

The highway here shows signs of its secondary status. When Stewart passed through, the pavement was new and unblemished, with sharply defined, regular edges. Today, the highway's appearance implies only minimal maintenance, as does the jostling ride produced by the rough surface. Tar and gravel have been used to resurface the eastbound lane and, in places, the shoulder. The weedy bank with blooming dandelions rising above the left side of the road is in contrast to the mowed condition revealed in the earlier photo.

SIX-LANE HIGHWAY

Six-Lane Highway

DuPont Boulevard extends the length of the state of Delaware. It was intended by its designer, Coleman DuPont, to be the "highway of the future." Planned early in this century, the roadway was to have lanes for "trollies," horse-drawn vehicles, and pedestrians, was well as for motorized vehicles. The highway was completed in 1924, and ever since has been the major north–south artery through the state.

Here, about three miles south of Wilmington, Stewart in 1950 described DuPont Boulevard as the "highway of the past" because its design did not conform to then-modern standards. The right lane had been squeezed into the right-of-way, leaving too narrow a shoulder. As a result, the curbs and mailboxes had become dangerously close to the flow of traffic.

But Stewart's judgment may have been premature. The appearance of the highway today does not suggest an outmoded design; indeed, the road even in 1980 might be mistakenly assumed to be of recent vintage. True, the irregular shoulder is too narrow in some places, particularly toward the crest of the hill. But the unpaved area adjacent to the pavement is not cluttered with signs, posts, or other structures, and its mowed surface may serve some the functions of a more conventional shoulder. The line of power poles is set well back from the highway, and even the mailboxes have been removed. The pavement surface is seemingly smooth and free of chuckholes or rough patching. A left-turn lane has been added at the crossover halfway up the rise, and the width of the traffic lanes does not appear to bring side-by-side vehicles too close together. The wide dividing median is a design feature that exceeds the minimum standards on some much newer stretches of U.S. 40 and even on the Interstate Highway System.

In Stewart's day, the daily traffic at this point, 22,688 vehicles, was the greatest of any stretch of U.S. 40. One-half mile beyond the top of the hill in the background, the traffic divided evenly between those vehicles heading south on U.S. 13, DuPont Boulevard, and those destined for Baltimore on U.S. 40. Today, the daily traffic flow here is about the same, 21,510, and about 14,000 use U.S. 40 to cross into Maryland. These facts, however, misrepresent the great increase in the number of vehicles bound for Maryland since 1950; nearly 33,000 vehicles use Interstate 95, which connects Wilmington with Baltimore by paralleling U.S. 40 to the west. The six lanes in the 1980 photo carry forty-two vehicles (Stewart commented that the twelve vehicles present in his picture represented heavy traffic), but the highway seems not to be uncomfortably or unsafely congested.

This section of U.S. 40 is no longer, as it was in Stewart's day, "the only part . . . that . . . attains six-lane width" outside of urban areas. Stretches with three lanes in one direction are common along the route today, even in remote and sparsely settled areas where a third lane is often added to two continuing lanes in order to facilitate passing on steep grades. The widest segment of highway along old Route 40 in 1980, incidentally, is just northeast of Sacramento, California, where six lanes are available in *each* direction.

BALTIMORE ROWS

Baltimore Rows

"The American city, east and west, is highly standardized," lamented Stewart in his introduction, "and a picture of one city stands pretty well for a picture of another." Thus, although in 1950 U.S. 40 passed through eight of the thirty largest cities in the United States, Stewart chose to include urban street scenes of only two of them, San Francisco, at the western terminus of the highway, and Baltimore, shown here at Fayette Avenue near Carrollton Street.

Steward did not choose this scene *despite* standardization. Rather, he selected it *because* of the standardization it shows—not of cities, but of Baltimore "row" houses. This monotony characteristic of older urban centers near the eastern seaboard prompted the self-consciously tactful comment about Baltimore in the American Guide Series book on Maryland: "Baltimore may be an ugly city; nevertheless it is charmingly picturesque in its ugliness."[3] Stewart observed more objectively: "Baltimoreans take [row houses] for granted. Visitors are amused or horrified."

Stewart, however, unequivocally admired the row houses in his photo: "Much can be said for the warm reds of their honest brickwork, for their simple doorways . . . and for the generally fine proportions of their facades. Moreover, they represent a real architectural tradition, developing out of a way of life that endured with much stability and homogeneity over several generations. We may contrast the more recently built districts of many cities where a dozen rootless imported styles of imported architecture clash in the same block."

The 1980 view displays that "clashing" of styles which Stewart deplored. The row houses on the south side of the street remain, but those on the north side have been torn down and replaced with the high-rise rectangular box of a public housing project. Contributing further to the variety of architectural styles is the complex of modern design in the background, which partially blocks our view of the old tall buildings in the heart of the downtown area.

Stewart also speculated that this neighborhood in 1950 was "on the downward path," as evidenced by "the littered street and the garbage can." While he identified features that were only superficial, his assessment of the trend of the neighborhood was correct, judging by conditions in 1980. The first two buildings, reached from the street by the closest pair of steps, are abandoned, gutted, and without windowpanes. Accumulations of bottles, cans, and scrap paper cover the basements and the first floors. The notice placed by a heating-plumbing contractor suggests that internal reconstruction is under way, although we could see no evidence of such work inside. Weeds are growing between the steps and within cracks in the sidewalk, and a small ailanthus tree, rooted in the stairwell to the basement, emerges above the sidewalk. The walls of these abandoned buildings, however, seem solid, and free of holes or fractures. Even the marble steps, which Stewart described as "scrubbed every morning," appear as clean and firm as they were in 1950, even though they no longer enjoy daily washing.

The second pair of steps leads to two buildings which are inhabited. From our present vantage the buildings look much the same as the closer structures, although the doorside porch lights, the stair railings, and the light color of the recently washed front brick wall suggest that people live within. A closer look would reveal curtained windows with potted begonias on the inside sills to confirm that suggestion.

Beyond the residences, the building with arched doorways and buttressed walls is a church, as it was in Stewart's day. After considerable driving amid the Baltimore rows, in fact, we were convinced by this landmark that we had finally found the photo site. Still farther down the avenue other row houses have been replaced with a low modern building.

Fayette Avenue in 1980 is no longer a major urban thoroughfare, as suggested by the lack of traffic and by the parking along the curb, which, in Stewart's time,was not allowed. The neighborhood through which it passes seems as quiet and undistinguished in 1980 as in 1950, even though the changes in the physical landscape are considerable. As it did for Stewart, the view for us may be taken as typical of a certain sort of urban scene: an old residential area in the inner city with vacant and gutted homes, tall unadorned project housing, and, in early morning hours at least, streets empty of moving vehicles or pedestrians.

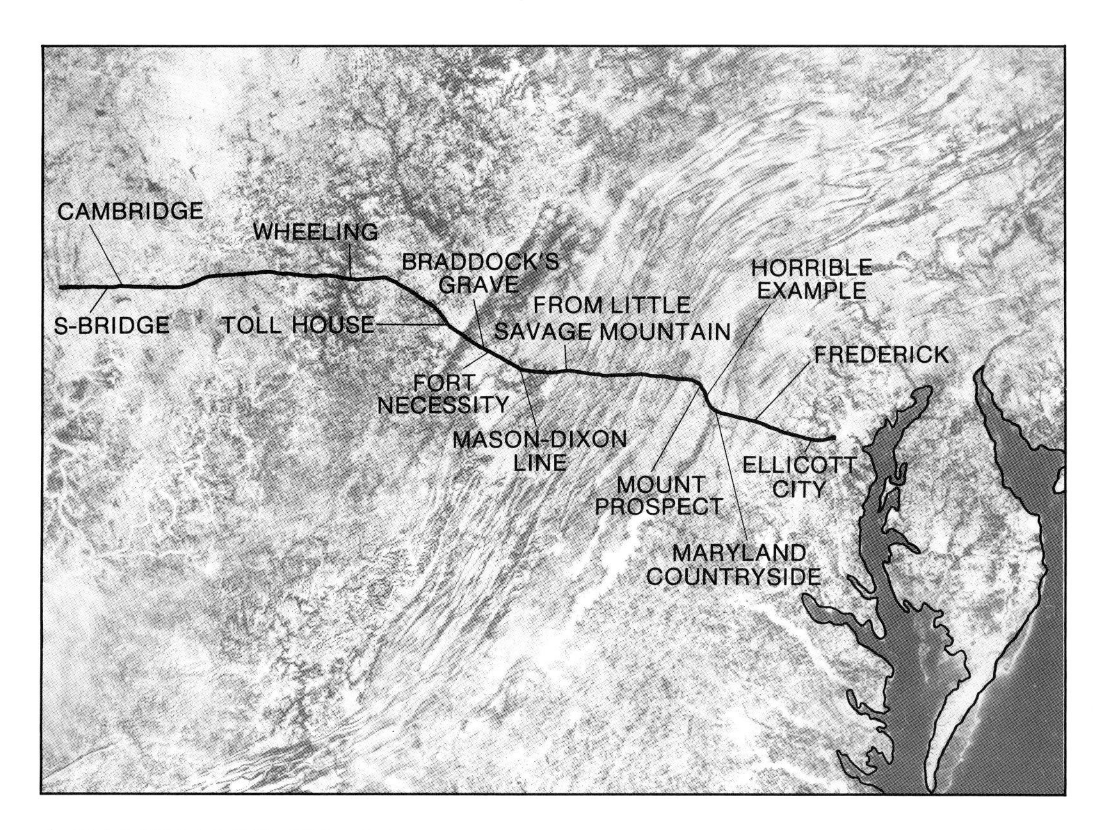

BALTIMORE TO COLUMBUS: APPALACHIAN HIGHLANDS

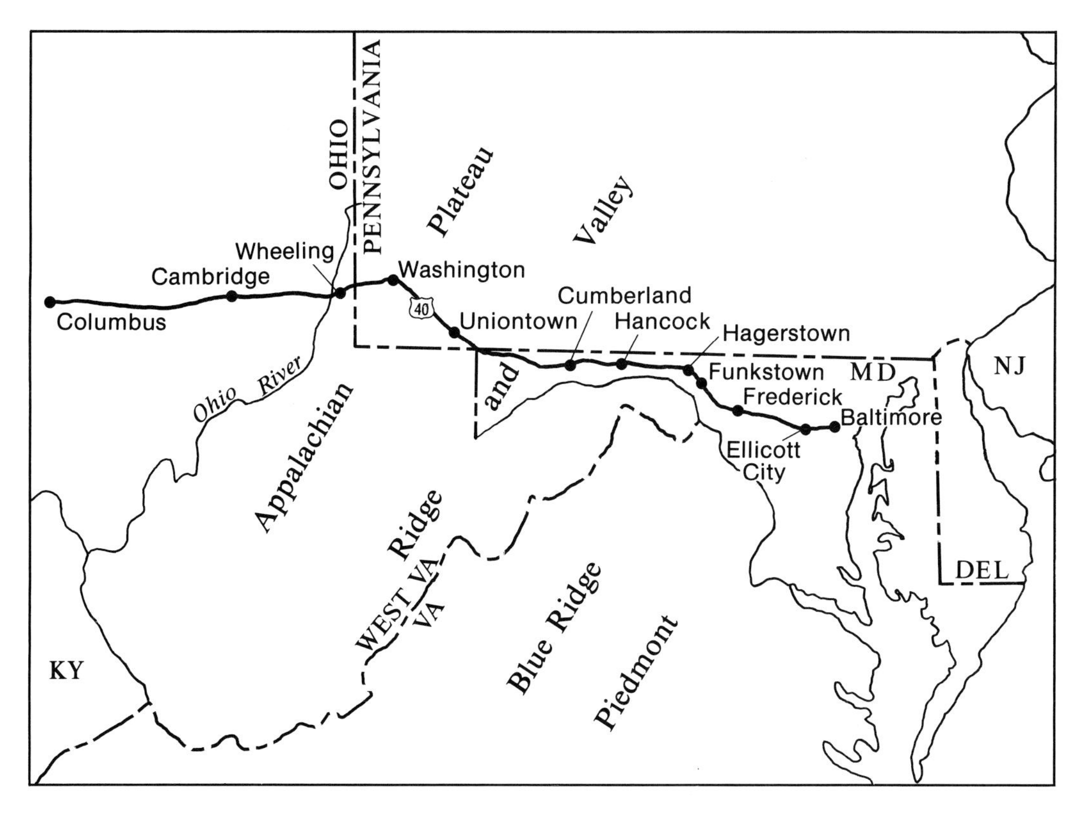

Baltimore to Columbus: Appalachian Highlands

FROM BALTIMORE, MARYLAND, to Columbus, Ohio, a distance of about 400 miles, U.S. 40 crosses the mountainous region of the eastern states. As far west as Frederick, it lies within the gently rolling terrain of the Piedmont, but then it traverses the more rugged topography of the Blue Ridge, the Ridge and Valley region, and, beyond Uniontown, the Appalachian Plateau. Several sorts of images characterize this region: the bucolic landscapes with forests on long ridges and fields in broad valleys near Frederick and Hagerstown; the high relief of steep wooded slopes and narrow canyons near Cumberland; the dissected lands between Uniontown and Columbus which Stewart described as "a maze of green hills . . . lacking in structure . . . partially farmed, partially wooded, here and there blackened by the dumps of coal-mines"; and the deeply cut valley bottoms of the Monongahela and Ohio Rivers. Most of this region shares the characteristics of scattered settlement, abundant woodland, and stream-carved topography.

U.S. 40 in 1950 was a major highway across this sometimes rugged terrain. Today, the route is a mixture of freeway and two-lane road, in places on the interstate system and elsewhere not, mostly on its early alignment but occasionally relocated. It is usually busy with traffic, but in a few stretches where a bigger roadway siphons off the vehicles it is surprisingly little traveled. As far west as Hancock, Maryland, U.S. 40 and Interstate 70 run together, with the U.S. route number used for the roads through the towns. At Hancock, the interstate turns north, and U.S. 40 continues west across the panhandle of Maryland before veering northwestward into Pennsylvania, where it rejoins Interstate 70 at Washington. Thereafter, into Columbus and all the way to Denver, Colorado, the two numbers remain either together or generally parallel. U.S. 40 is often the more local route that provides access to towns and farms, while the interstate carries most of the through traffic.

This segment of the highway is perhaps the most important historically. From Cumberland to Wheeling, West Virginia, U.S. 40 is on the route of the original National Road. Authorized by Congress in 1806 and completed by 1818, this publicly financed road was "a notable achievement. . . . A strip 66 feet wide, cleared of trees and underbrush. A leveled roadbed thirty feet wide. Grades and curves moderate, so moderate indeed that many of them still remain on the present highway. Twenty feet of the surface covered with broken stone from a depth of twelve to eighteen inches." The project was a critical event in American history: it set a precedent for the use of federal money for "internal improvements," it gave Baltimore an economic hinterland which encouraged its emergence as a major shipping port, it allowed the settlement and development of the lands beyond the mountains in the Ohio Valley, and "in one way and another it may have done much to preserve the United States as one nation." It is not surprising that Stewart devoted so much attention to features of history on this segment of U.S. 40.

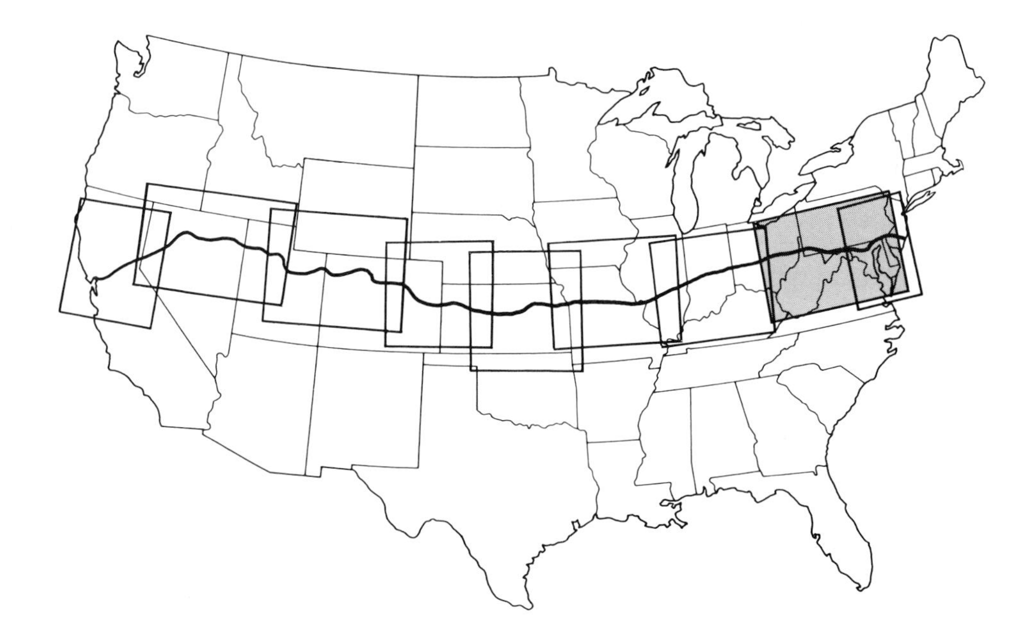

ELLICOTT CITY

Ellicott City

Ellicott City in 1950 was the county seat and commercial center for rural Howard County, just beyond the western fringe of urban Baltimore. Born as a Quaker mill town and stimulated later as a station on the westward-extending B & O Railroad, Ellicott City had been a commercial center for nearly 300 years.

The businesses along its narrow Main Street within the view of Stewart's photo included several lunch counters, a taxi company, a laundry, and stores selling food, shoes, liquor, and general merchandise. Workingmen, often leisurely perched on the front steps of the rooming houses where they lived, were not the only customers along this busy street. Families from around the county came into town for shopping, especially on Saturday evenings when they were lured by the late hours of such department stores as Caplan's, located half a block up on the left. The streetcars carried people into town (and, at one time, out of town all the way to Baltimore), unless, as often happened, autos parked too far from the curb blocked their progress. A resident who was a teenager in the years before Stewart passed through recalled recently the excitement associated with the influx of shoppers: "We would sit in the wooden chairs we carried from our homes to the pavement, watching the passing parade with amusement. . . . We exchanged stares with the country people . . . the men in bib overalls, the women in out of date dresses with a brood of children clinging to the irregular skirt."[4]

Ellicott City's Main Street in 1980, now Maryland State Highway 144 instead of U.S. 40, appears at first glance much the same as it did in 1950, despite the conspicuous absence of streetcars. The appearance of the street, with its narrow winding pavement and its border of multistoried rooming houses built of cut rock, still suggests both its great age and its industrial origins. But the human activities along it are in part changed. The downtown area now appeals as much to the out-of-town tourist, attracted by the new railroad museum, the old buildings, and the specialty shops, as to the more conventional local shopper. Along the street today we find restaurants offering seafood blintzes and French wine rather than ham sandwiches and coffee, shops selling woven baskets and imported fashions rather than surplus military goods and used clothing, and male pedestrians in brightly colored shorts and hats rather than dull work pants and baseball caps. Although more traditional stores remain, this lower stretch of Main Street is a tourist street.

Wishing to maintain the tourist appeal of its "historic" appearance, the downtown area organized in 1974 as the Ellicott City Historic District. The District's Commission has the right to review proposals for construction, and it attempts to maintain the "plain look" of the downtown buildings, with their characteristically Quaker lack of embellishment, their row-house design, and their use of local construction material. The recent photo illustrates several architectural features encouraged by the Commission: canvas (rather than metal) awnings, shingled door and window hoods, shuttered windows, and small store signs flush against exterior walls. Paint has been stripped from the old Colonial Inn and Opera House to reveal the natural yellow hues of granite quarried nearby. If the present goal of maintaining the old face of Ellicott City is achieved, some current resident may be able in the future to look back to 1980 and observe, as an old-timer recently did: "Surrounded by the homes of suburbia, encircled by cars racing on expressways . . . [Ellicott City] has retired behind her granite walls and kept her wry personality . . . like an old lady who refuses to sell and lives with her memories though skyscrapers peer into her garden."[5]

FREDERICK

FREDERICK

OUR VIEW ON Patrick Street in downtown Frederick reveals a group of elderly people who have just finished the guided tour of the Barbara Frietchie House. They are either resting in the shade on this hot sunny afternoon or are taking a minute to look at Carroll Creek and the small wooded park beyond. Their bus, appropriately parked on a stretch of roadway designated for tourist parking only, will soon carry them off to another spot in the historic county seat of Frederick County, Maryland.

"One has a little difficulty," Stewart commented, "in saying just what the museum represents." It does not commemorate a person or event that shaped society, and the building which houses it did not exist in the Civil War days of Barbara Frietchie. The poem by John Greenleaf Whittier which made the name Barbara Frietchie famous most likely depicted, in fact, a fictitious incident.[6] Still, the image of a ninety-year-old woman hanging out of the upper window and defiantly waving the Union flag while a Confederate army marched in the street below appeals to American ideals of patriotism, dedication, and courage.

The house is virtually unchanged since Stewart's time. The west-facing brick wall is unblemished, the roof appears sound, and the trim is neatly painted and maintained. The sign announcing the house is different, the newer one being more modest and less cumbersome, but the structure supporting the sign is the same. Not visible from this vantage point is the newly installed sidewalk of small hexagonal bricks in front of the house. The windows, conspicuously open in 1950, are tightly closed in 1980, suggesting that today air conditioning cools the interior. Curiously, no flag hangs from the upper window in the recent photo.

If Barbara Frietchie were to lean "far out on the window sill" today, she would see many buildings that have been standing for over a hundred years. Yet, changes have occurred even over the last thirty years. As in many historic areas of major cities, the nineteenth-century buildings along the street have been refurbished as professional offices. Lawyers and insurance agents occupy much of the former residential space. The interior of the first building beyond the Frietchie house is being rebuilt for such use, and the bed of the truck parked on the sidewalk is filled with the rubble of old inside furnishings. Farther away, where a parking lot borders the sidewalk, Delphey's Sporting Goods store is set back from the street, as it was in Stewart's time. Then, Harley-Davidson motorcycles were boldly advertised with a large sign, but today, appropriately for 1980, bicycles and a single "Moped" adorn the show window. Beyond Delphey's, at the corner, a large open area has been cleared of old buildings, and a new brick County Courthouse of modern design is nearly completed, although it is hidden from our view. Finally, the sign of the Francis Scott Key Hotel is gone, signifying the new function of the hotel: it has become the Homewood Retirement Center.

Other features in this landscape also reflect change. The power pole with its three wires on the cross arm seems identical in both photos, but the clutter of wires is much more pronounced in 1980 than in 1950. Massive street lighting units are crowded together above the roadway, whereas the only streetlight visible in Stewart's photo is a single delicate pillar and globe on the bridge. The television antennas, which Stewart said were erected between mid-1949 and mid-1950, are gone. The sidewalks and curbs have become weedy, and at the base of the downspout in the second building beyond the Frietchie House a crack in the concrete has allowed a snapdragon to grow up and bloom vigorously next to the wall. Pigeons, attracted by the cornices, overhangs, and broad windowsills of the old buildings, glide across the sky, and give a touch of the typical urban American scene to unusually historic Frederick.

MARYLAND COUNTRYSIDE

THE VIEW OVER Middletown Valley was ideally clear and bucolic when Stewart took his picture, but in the recent photo the view seems marred by the hazy atmosphere and the sprawling subdivision. It would be unfair, however, to interpret either of these dramatic differences in solely negative terms.

The haze should not be confused with smog or other pollutions produced by human activities. It is composed of water droplets, and consequently it may be accurately described as a thin fog. Such hazy conditions are far more common in the humid East than in the arid and semi-arid West. When summer clouds develop into thunderstorms east of the Mississippi River, the haze caused by the associated high humidity obscures the edges of clouds, and restricts visibility. Skies are thus visually muted, even while they produce intense rainfall and spectacular thunder and lightning. In the western plains, mountains, and deserts, on the other hand, lower humidities mean less haze, and towering cumulus clouds are etched sharply in a brilliant sky. It is not surprising, then, that a reader wanting to see photographs of beautiful clouds picks up *Arizona Highways* rather than a pictorial magazine for country east of the Mississippi.

We had waited for several days, hoping that a change in the weather would allow us to take this and several other nearby photos with a clearer sky. But even after the passage of a cold front, typically followed by cooler and drier weather, we continued to look out over misty landscapes. Stewart took his picture in early fall, a less humid season.

Unlike the haze, the subdivision is obviously the result of human activities. The people who live in these homes probably commute to Frederick, which lies only a few miles to the east. The houses are laid out on mostly straight segments of street, but the segments are joined by sweeping curves. From our vantage point, this design of the roadways and the varying angles of the orientation of the buildings produce a subdivision pattern that is not rigidly rectangular. Taken together with the irregular field boundaries which Stewart admired, and the combination of straight and curved borders produced by contour plowing, the scene, even with a subdivision, is not displeasing. The loss of productive cropland prompts many people to view such development as unfortunate or "wrong," but in this case we find the sturdy white homes scattered in a pleasing pattern on the rolling countryside to be rather attractive.

Most of Middletown Valley continues to serve as agricultural land, and crops of corn and alfalfa are dispersed among pastures with grazing dairy cattle. The growth of trees and shrubs along the field boundaries, obvious on the fence line in the foreground, suggested to Stewart the "rapidity with which this whole eastern part of the United States would return to forest if the hand of man were removed." Birds contribute to this spread of woody plants by eating and subsequently dispersing seeds of such plants as wild raspberry and honeysuckle. Catbirds, robins, and mockingbirds were busy in the thicket of berry bushes immediately below us on that late July morning while we sat awaiting in vain some change in the weather that would allow us to see beyond the subdivision and to view the landscape described by Whittier as "fair as the garden of the Lord."[7]

HORRIBLE EXAMPLE

Horrible Example

Stewart hated billboards. "The large billboards . . . are always placed in the most conspicuous spots, and have designs and colors carefully chosen to clash with the background . . . They are an abomination!" Yet, Stewart did not view with disdain another ubiquitous element of the American landscape, the utility pole: "Pole lines and wires may be accepted. They do their work without striving to be conspicuous, and often their not-ungraceful curves add a touch of interest, an intricacy of pattern, even some beauty."

In this view at the western edge of the small town of Funkstown, both billboards and utility lines flank Alternate U.S. 40 as it crosses historic Antietam Creek on a 150-year-old bridge. Stewart found much to admire in the scene, and yet he felt that it was desecrated: "The billboards ruin everything. The historic flavor, the old-time architecture, even the beauty of the wooded hillside—all are sacrificed." The four billboards in 1980 would please him no more than the three in 1950. The lengths of the structures seem shorter than in 1950, and they all lack the fringe of lattice which formerly adorned their bases. The subjects advertised have also changed, but only one suggests the nature of the times in 1980, proclaiming that a "power roof vent cuts cooling costs." Stewart would, at least, admire the alliteration.

Whereas Stewart observed that the billboards "blast themselves into the viewer's consciousness," he described the utility lines as "inconspicuous" except against the stark sky. Today he might find the additional wires and poles to be as objectionable as the billboards. Four telephone cables cross the creek on the far side of the bridge, a new pole with eight power lines crosses the near side, and two streetlights mounted on their own wooden poles connect to the power line with their own wires. This clutter of wires is not reminiscent of the utility poles marching across the sweeping grasslands of western Kansas or sagebrush of central Nevada where Stewart could rightly admire the "majestic" pole lines.

The small piece of cardboard in the foreground announces in hand lettering a yard sale up the side street to the right of the photo. Stewart could accept such signs because they seemed to have "a certain rooting in the soil. . . . One feels a difference between a home-produced: 'Stop at Joe's Service Station for Gas—Two Miles,' or 'The Liberty Cafe—Short Orders at All Hours—Give Us a Try!' and some gigantic rectangle advertising tires or beer." Alas, Stewart would find little comfort in knowing that present federal regulations permit only signs advertising local motels, service stations, cafes, and other businesses along the Interstate Highway System, for in order to catch the attention of the mile-a-minute tourist, these signs have acquired all the characteristics of those which he disdained as "usually advertising the products of mass production."

Indeed, the billboards of Antietam Creek distract our eyes from other features in the landscape. The line of small pines at the edge of the drop-off into the creek channel fringes the neatly mowed grass on the terrace below the highway and marks the edge of the grounds of a new business just to the left of the photo view. The short, slender streetlight pillars on the bridge in 1950 are gone, having been replaced by the two modern light standards. The sign identifying Antietam Creek continues to be located on the wrong side of the road, since the tall concrete wall which holds the road fill of the bridge approach extends farther away from the bridge on the opposite side of the street and makes difficult the anchoring of a sign post. The trees in the background seem to have grown in density and height, although the tall individual tree behind the billboards on the left still is alone in breaking the sky line.

MOUNT PROSPECT

Mount Prospect

"As handsome a historic structure as anywhere faces the three-thousand-mile course of U.S. 40," Stewart said of the 200-year-old Rochester house, called Mount Prospect, on West Washington Street in Hagerstown. The neighborhood was described in the American Guide Series book on Maryland as having "houses of Georgian Colonial and Classic Revival design set far back on spacious lawns behind the trees for which the city is notable. . . . All have the trim well-cared-for look seen only in places where the majority of the inhabitants own their own homes and are proud of them. [Residents have an] appreciation of growing things and of the natural beauty of the valley."[8] Mount Prospect was one of six buildings identified in that book as having historical significance in Hagerstown.

Today, only the presence of a parking meter, albeit of more modern design, and the gradual slope of the street bear any resemblance to the scene portrayed in Stewart's photo. The gentle rise upon which the Rochester house was built, and from which the name Mount Prospect was derived, is now accentuated by two tiers of bare concrete walls. The stone stairway, the brick sidewalk, and the stone curb, all of which Stewart admired as "in the early American tradition," have been replaced by the modern American tradition of unfinished concrete. The "Victorian touches" of the graceful iron fence and the gate with "elaborately designed gateposts" in the 1950 scene contrast strongly with today's practical and stark steel chain-link fencing. In place of a canopy of vine-covered trees arching over lawn and flower beds, a few light standards stare down on pavement and parking meters.

Only five years after Stewart passed through Hagerstown and took his photo, the city government, which then owned the building, demolished Mount Prospect and replaced it with the parking lot. We found the site easily, having a complete address, but we were understandably skeptical. A local mailman confirmed for us the former presence of Mount Prospect on the lot, and recalled that the family which had owned the house felt it was too expensive to maintain and sold it to the city, which was eager to increase parking space only two blocks from downtown Hagerstown. A spokesman for the city explained the justification for the demolition: "Funding was not available for needed extensive renovations to the building which, of course, was not producing any tax revenues."[9]

The Rochester house would likely still be standing if it had been located down South Prospect Street, the side street visible on the extreme left side of the recent photograph. Many homes on South Prospect are listed on the National Register of Historic Sites, and the street is within a National Register District. Some of these houses are elegant, but none merits as fully as did Mount Prospect the high praise of the neighborhood in the American Guide Series book.

Of the nearly one hundred scenes in Stewart's work, none has changed so completely or has had such an alteration of personality as Mount Prospect. In general, we found changes in the appearance of the landscapes along U.S. 40 to be a source of enjoyable fascination, sometimes even when the changes were the sorts commonly decried. We have just viewed with some appreciation the new patterns created by the transformation of Middletown Valley's agricultural fields and pastures into a housing development. In a later photograph, at the eastern fringe of the town of Richmond, Indiana, we will express interest in the forms and features in the commercial strip development which has replaced expanses of corn. Still farther west, in the California Sierra Nevada, we will admire the gleaming white lanes of a new freeway. But at Mount Prospect, more than at any other site along U.S. 40, we regret what has happened.

FROM LITTLE SAVAGE MOUNTAIN

From Little Savage Mountain

Westward from near the summit of Little Savage Mountain, the rolling upland of the Allegheny Plateau extends to the horizon. The most distant ridge visible in Stewart's photo is lost in the haze of the more recent view, but in both pictures forest seems to cover the ridges whereas fields appear to stretch across the broad gentle uplands. Actually, topographic maps of this area showing boundaries between forests and clearing for pasture and cropland suggest that the plant cover is not strongly related to topography. In this view, for example, the field beyond the overpass and facing the photo site is on a slope no more gentle than the wooded slope facing the opposite direction on the other side of the hilltop, but hidden from our view.

The absence of strong relationships between land use and topography today reflects, in part, the steady abandonment of marginal farmland in Appalachia, forced by economic considerations over at least the last forty years. In Garrett County, where this scene is located, over half of the land area was in farms when Stewart drove along U.S. 40, but today that fraction has shrunk to one-third. The rapid invasion by trees of lands no longer regularly plowed is suggested by the thicket of sumac and locust which has sprung up in the vacant land on either side of the highway in the foreground.

The condition of the buildings of the farm also hints at the decline in agricultural activity in Appalachia. In 1950, the farmstead was neatly maintained with board fences and sturdy buildings, all shining with fresh white paint. Today, the farm appears shabby, with patches on the barn roofs, dull and darkened exterior walls of most buildings, and an absence of fences setting off the fields or bordering the highway. The old orchard behind the farmhouse which Stewart commented had "been allowed to deteriorate . . . until the actual pattern of an orchard is scarcely distinguishable" is now virtually gone. The patch of trees and shrubs beyond the old orchard site is curiously still present and little changed, perhaps reflecting its presence on ground too rocky to plow. The ornamental trees in the yards adjacent to the houses have grown gracefully. The strips of corn and fallow on the hillslope, carefully arranged on the contour, indicate that the farm is not abandoned, although it certainly is not "given over to stock-raising, chiefly horses," as it was in Stewart's time.

Even more than the farm buildings or the farm fields, the highway has changed dramatically over the last thirty years. The single two-lane roadway has given way to a confusing pattern created by the old road, a parallel freeway, a relocated side highway, and access ramps connecting the major through routes. The barns which stood proudly beside U.S. 40 in 1950 now are wedged tightly between two roadways and close to a third. The dominance of the expanded highway complex is in contrast to the situation a century ago, when the then infant railroads usurped the traffic of the flourishing National Road, as described by Robert Alberts in his booklet *Mount Washington Tavern:* "Weeds began to grow where the stages and wagons had rumbled. Landowners began to encroach on the Road, moving out their fences until it had shrunk from 66 to 40 or even 30 feet."[10] Today a strip of land 700 feet wide serves as the transportation corridor for the highway.

The new alignment of the National Road over Little Savage Mountain necessitated by the adjacent freeway made our photo site slightly different from Stewart's. The old road was at a higher elevation and generally followed the edge of the trees beyond the highway in the foreground. Just to the left of this section of the photo, we found a remnant of the former highway on the hilltop, where thirteen garbage bins, located conveniently at the highway junction for rural trash collection, were lined up on the old broken pavement.

MASON-DIXON LINE

Mason-Dixon Line

After dropping gently down into a valley, U.S. 40 climbs rather sharply up the distant ridge in a broad sweeping curve. Halfway up the hill, where the pavement changes from asphalt to concrete, the highway passes from Maryland to Pennsylvania. "The surface of the highway . . . gives visible and concrete reality to one of the most famous abstractions in American history," proclaimed Stewart. "This is the Mason-Dixon Line." The division between slave and free territory before the Civil War, this political boundary today is but another line separating states. A steel structure cantilevered over the highway supports a large sign welcoming travelers to Pennsylvania. Newcomers to Maryland are greeted by no fewer than seven signs strung out from the state line toward the photo site. These announce such highway regulations as speed limits, fines for littering, and prohibitions against passing stopped school buses, and offer such information as the route number and the designation of the roadway as a "snow emergency route."

The recent photo is taken from a spot a little closer to the state line than where Stewart stood, because vegetation in 1980 blocked the view from the original picture location. The growth of woody plants, in fact, is the most conspicuous change illustrated by the photos. On page 55 of Robert Bruce's *National Road,* published in 1916, a photo of this swale shows open pastures bordering both sides of the roadway.[11] By 1950, Stewart noted that "the trees [had] started to take over." He anticipated the further establishment of trees and shrubs on the road cut in the distance: "Such a cut will grow over in the course of a few years, and no longer be so sharply distinguishable from the rest of the landscape. In the eastern part of the United States, if roadwork ceased for as long as ten years, the raw cut would largely disappear from the landscape. A growth of bushes and saplings is already making headway on this cut."

By 1980 the advance by woody plants is complete. Even the old orchard has been engulfed. As noted in the view from Little Savage Mountain, such abandonment of agricultural land typifies Appalachia.

Yet, the encroachment of forest upon farmland in the region is balanced in part by the reverse process. The geographer John Fraser Hart suggests that in the southern Appalachians a kind of rotation in land use is practiced, with farmers clearing and planting small agricultural fields, then abandoning them to pasture before forest regains dominance: "After a generation or so under trees the soil [has] regained enough of its fertility to be worth reclearing, and the cycle of cultivation, pasture, and woodland [is] repeated." This process tends to make the landscapes involved look "disorganized and patternless, unkempt, untidy, and untended," even though it reflects a modest and conserving attitude toward land use.[12]

The small settlement of Strawn astride the state line had nearly a dozen buildings in 1916, as evidenced by the *National Road* photograph. On the right side of the highway just inside Maryland, we found two collapsed wooden frame buildings, one with the word *groceries* still discernible though badly faded. Across the road, however, the State Line Methodist Church has been painted, and adorned with a new roof. But the barn behind and upslope from the church is gone, having been replaced by a plantation of pines. The home beside the highway and beyond the church must have been hidden within the trees in Stewart's photo, for it is obviously not of contemporary design. In the foreground, as indicated by the mailboxes, two new houses have been built, set back in the protection of the trees.

FORT NECESSITY

Fort Necessity

About nineteen miles northwest of the Maryland state line, U.S. 40 follows the route of the National Road, crossing a gentle ridge called Mount Washington. The highway is not visible in either photo, but its alignment is just beyond the large building in view on the skyline in Stewart's picture, where it is gradually ascending from right to left. The valley from which the photos were taken is called Great Meadows and is a unit of the National Park System, Fort Necessity National Battlefield.

Fort Necessity is the site of a military skirmish in 1754 between Britain and France, which were vying for control of the Ohio River area. A young British lieutenant colonel named George Washington had led his colonial force westward from Virginia. He established an operating base in this wet meadow on May 24, 1754, while his men improved the trail over the mountains "to make the road sufficiently good for the heaviest artillery to pass." Although at first selected as temporary headquarters, the site was soon to be tested as the "charming field for an encounter" which Washington initially recognized it to be, with its natural entrenchments carved by streams.[13] The French, who had already established settlements farther west, sent forces to resist the British incursion. On July 3, with heavy rain falling, 700 French and Indian forces attacked the 400 British colonials from a forest then located where the group of planted trees appears in the photos. The fighting lasted but a day; a truce was negotiated. Washington led his forces back across the Appalachian Mountains, and the French burned the fort.

Fort Necessity was reconstructed in 1932 and became a unit of the National Park System a year later. That first rebuilding of the stockade is what Stewart saw and photographed. The design of the reconstructed fort was based on early maps, drawings, and descriptions, but these sources were not free of ambiguity. For the first half of this century, historians argued about whether or not the fort had been in the shape of a triangle or a diamond. They all shared the assumption, however, that the vertical logs of the wall were emplaced in a low ridge of earth piled up to increase the height of the structure, as is seen in Stewart's photo.

A few years after Stewart passed through Fort Necessity, exploratory excavations of the ground within the picture view unearthed the bases of the logs of the original walls, preserved in the saturated soil. Further digging led to the discovery that the stockade had been neither triangular nor diamond-shaped, but circular. Moreover, it was much smaller than earlier thought, and had been located within the low embankment of earth which had served as an outer parapet. With an obviously inaccurate reconstruction on its hands, the National Park Service dutifully tore down the fort and meticulously reconstructed it according to the new information. It was this second version of Fort Necessity which we photographed in 1980.

According to David Lowenthal, a geographer and scholarly commentator on the American scene, Americans worry little about the authenticity of historical reconstruction as long as the reconstructed structures evoke an aura of the distant past.[14] The experience at Fort Necessity argues against this idea. On the other hand, the setting of the fort is clearly not identical to that of 1754, a fact which Stewart also recognized. Furthermore, the trees in the background (saplings in Stewart's photo) were planted to simulate the original forest, but some of these plantings included stands of pine, even though such trees were not present in this area in the eighteenth century. The neatly mowed sod does not duplicate the original meadow grasses, which probably more closely resembled the tall grass apparent beyond the fort in the 1980 photo. The paved walk is obviously a strictly modern artifact.

The battlefield site has been a public memorial for nearly fifty years. George Washington himself held considerable affection for the spot, and in 1769 he bought 235 acres which included the meadows and remnants of his stockade. Washington felt that the tract, bisected by a route that would become the National Road, was "an excellent stand for an innkeeper," and that its meadowlands were "fit for the scythe."[15] A tavern, prominently visible on the hill in Stewart's photo but hidden behind the trees in the recent view, was built in 1828, and now is a museum in the National Battlefield. Today, the meadow grasses may not be harvested for hay, but they are cut regularly by power mowers. Undoubtedly, the Park Service has done its best to ensure that many a twentieth-century visitor may find the Great Meadows a "charming field" for an encounter with the nation's past.

BRADDOCK'S GRAVE

Braddock's Grave

About a mile northwest of Fort Necessity, U.S. 40 skirts Braddock's Park, a twenty-three-acre reserve administered as part of the Fort Necessity National Battlefield. It memorializes the death of the English general Edward Braddock, who was fatally wounded in a later battle of the same war commemorated at Fort Necessity.

In 1755, the British forces, after again being defeated, this time near present-day Pittsburgh, were retreating eastward when the general succumbed to his injury. In order to avoid the detection and possible disturbance of his body by the enemy, his men buried him within the crude road that they had earlier improved for military reasons, at a spot down the ravine to the right of the photo view. His remains were unearthed during work on the roadbed in 1804 and reinterred on this flat above the ravine, but the handsome granite monument was not erected until 1913. The park land surrounding the grave site and monument was purchased by local citizens who eventually turned it over to the State of Pennsylvania. The National Park Service assumed responsibility in 1962.

The appearance of Braddock's Park is sharply different in the two photos. In 1950, the grounds appeared unkempt, with weeds as well as litter strewn through the grass. Braddock's Road was indiscernible among the dirt roads which unceremoniously crisscrossed the memorial. The monument itself seemed to shrink beneath the sheltering limbs of a towering pine. Stewart commented on the mood of the place when he photographed it: "The picture itself is not inharmonious with the mournful and inglorious death—the dark tree, the hills muted by mist, the low-hanging gray cloud, the crucifix-like pole and cross arm."

In the 1980 photo, not only the clear skies and bright sunshine contribute to a more cheerful landscape. The planted sod is neatly mowed, and the access drive to a parking lot just to the right is clearly marked by its gravel surface. The trace of Braddock's Road is marked by a path covered by chipped wood and bark crossing the grass; a low wooden sign announces simply, "Braddock's Road." The Braddock monument plaque emplaced on the irregular rock at the side of Stewart's photo has been moved to a position farther to the right, and the granite monument emerges into the sunlight to dominate the scene. A large wooden sign facing the highway identifies the park, and an aluminum informational sign explains the story of Braddock and his demise. The original "Braddock's Park" sign beside the highway remains in its 1950 position.

The neatness of the park in 1980 seems somewhat incongruous with what is memorialized, and the weedy overgrown look of the view in 1950 seems more appropriate. But, even more fundamentally, we wonder exactly what is commemorated. The bodily remains moved from the old road to the adjacent knoll were only "said to be" those of General Braddock. Moreover, Edward Braddock is hardly a military figure familiar to the contemporary American. The incident at Fort Necessity at least involved a person who would become noteworthy in the American Revolution.

Recent archaeology at Fort Necessity resulted from an unexplainable hunger for historical facts; the writer of the National Park Service publication concerning the project observed, "Even the most conscientious scholar would probably admit that the precise location and appearance of Fort Necessity is of no great consequence to American history,"[16] yet conscience apparently dictated a major reconstruction to conform to the newly recognized "facts." If future archaeological findings should reveal that Braddock's true remains rest elsewhere, would a similar twinge of conscience result in relocating the memorial? Or are we here commemorating something more abstract than an individual's life and death—the complex of associations that are attached to an entire epoch of our history, perhaps, and of which Braddock's death is only emblematic?

TOLL HOUSE

Toll House

Two miles west of Uniontown, U.S. 40 rises and falls over the rolling and hilly country of the dissected Allegheny Plateau. Beside the highway to the south, the Searight Tollhouse sits on a hilltop with a good view in both directions. It was built in 1835 by the State of Pennsylvania, which had assumed responsibility for its portion of the National Road in that year.

In Stewart's time, the tollhouse was in poor condition: ". . . the building is in danger of falling to pieces. Much of the glass is gone. The porch-roof and the roof of the rear extension have both been broken. The porch posts at the front, where doubtless they were occasionally banged by wagon hubs, have gone out entirely." The overgrown weedy yard surrounding the house also suggested neglect.

By 1980, the building had been restored. Each of the windows has glass, and the large panes have been replaced by smaller ones more in keeping with the age of the structure. New roofs of wooden shingles adorn both the tower and the main building. The missing porch posts have been replaced, a railing added, and all of the wood painted white. The interior has also been restored with furnishings reminiscent of the time, and serves as a museum. A parking lot, with pavement painted green, is in back, with the entrance drive visible in the foreground. New informational signs and a wooden split-rail fence also distinguish the recent view from Stewart's. Curiously unlike either Fort Necessity or Braddock's Grave, but certainly more appropriate for the 1830s than a lawn, the yard remains weedy and unmowed.

The geographer David Lowenthal has criticized the tendency to isolate historical structures from the everyday world by keeping them on "green shaven lawns" behind the boundaries and entrance gates of parks. He quoted the critic John Ruskin, who evaluated the differences between historic preservation in England and on the European continent:

> Abroad, a building of the eighth or tenth century stands ruinous in the open street; the children play around it, and peasants heap their corn in it. . . . No one wonders at it, or thinks of it as separate, and of another time; we feel the ancient world to be a real thing and one with the new. . . . We, in England, have our new street, our new inn, our green shaven lawn, and our piece of ruin emergent from it—a mere specimen of the middle ages put on a bit of velvet carpet to be shown, which, for but its size, might as well be on the museum shelf.[17]

The Searight Tollhouse is set behind both a fence and a chain which blocks the parking lot entrance in off-hours, but otherwise it seems hardly so isolated that it "might as well be on the museum shelf." The road which it once served still functions at its doorstep as a roadway of both casual travel and commerce. Billboards facing the tollhouse accentuate these functions of the highway. Utility poles and wires carrying electricity and common telephone conversations break the skyline. And immediately adjacent to the tollhouse on its east side is a coal retailer, whose signs, appearing as dark rectangles just beyond the crest of the hill in the 1980 picture, advertise "run of the mine" and "lump" for twenty dollars a ton. Like continental Europe's "heaps of corn," the piles of coal are appropriate for the locale, and they serve to give Searight Tollhouse a real-world setting.

WHEELING

Wheeling

Photographs of the terrain along the Ohio River around Wheeling, West Virginia, indicate that the hills mostly lacked woody plants at the turn of the century. The once denuded Wheeling Hill, from which this picture was taken, was returning to woodland in Stewart's time, and was completely reforested in 1980. We spent much of an uncomfortably warm and humid afternoon unsuccessfully traversing the slope on narrow roads and overgrown walkways looking for the viewpoint. Whereas Stewart took his photograph "looking out across the rooftops," we were forced to accept a close approximation looking out *from* the rooftop of Lincoln School, near the base of the hillslope. The friendly youthful employee who accommodatingly led us to the access ladder expressed surprise that anyone would travel across the nation to photograph a city that "has only one or two streets with any action." To him at least, Wheeling is no longer the "bustling hub of the West," as a historian recently described the nineteenth-century city.[18]

Ebenezer Zane and his two brothers probably did not foresee the development of a "bustling" city when they first settled this spot at the junction of the Ohio River and Wheeling Creek in 1769. Almost from its beginnings, however, Wheeling has been a busy settlement, largely because of its importance as a transportation town. Boats and barges on the Ohio River were able to ply their ways upstream, in times of low water, at least this far. Wheeling was chosen as the western terminus of the original National Road which swung down Wheeling Hill to the right of the photo view and paralleled the river by following the street in the foreground. The B & O Railroad also reached the Ohio River here, a few miles to the south. It may not have been an exaggeration, then, when in 1916 Robert Bruce reported that "perhaps no place of its size in the United States has today as much through east and west travel as Wheeling."[19]

That distinction probably belongs elsewhere in 1980, but the most obvious changes revealed in the photos involve transportation. The new Fort Henry Bridge carries traffic following both Interstate 70 and U.S. 40, replacing the historic suspension bridge that had been built in 1856. The "Steel Bridge," visible in Stewart's photo just south of the suspension span, is gone, as is the urban trolley line which it supported. We see no boat on the river in 1980, although the Ohio is still a vital transportation waterway. Alterations in downtown Wheeling to accommodate automobiles—for example, the demolition of old buildings to make way for parking lots and a parking ramp—are typical of modern American cities.

"It is like a vision of Paradise," Ebenezer Zane told his friends of this spot, as reported in the American Guide Series book of West Virginia.[20] But unlike the Heavenly Paradise, which is usually imagined to be overhead, or at least somehow "up," the preferred lands in Wheeling were "down" on the rich, flat bottomlands that were easy to build upon and conveniently located for commerce along the river. The higher elevations on the steep, inaccessible hills were inhabited by persons with less money, and eventually these slopes were dotted with slum housing. As a result, Wheeling earlier in this century, according to a recent history of West Virginia, was a classic example of the West Virginia social geography in which "the people who lived in the bottom looked down on the people on top."[21] Today most of the affluent residents seem to have fled this paved Paradise to outlying neighborhoods.

CAMBRIDGE, OHIO

Cambridge, Ohio

About fifty miles west of Wheeling, U.S. 40 passes through Cambridge, Ohio. In the American Guide Series book on Ohio, Cambridge is described as "an old town stretched out along a high ridge overlooking rolling hills to the north and south. . . . Its tone is set by the old-fashioned courthouse with its loungers in the park. Few modern store fronts are seen, and many of the business structures are drab, stern creations of the 1870s and 1880s. But automobiles clatter through the town all day, and Cambridge is prosperous if not wealthy."[22]

Stewart's picture, taken on a drizzling fall morning, suggests an old, perhaps decaying, town in which the words "drab" and "stern" appropriately describe the buildings. Our photo, on the other hand, was taken at noon on a sunny summer day. The bright light makes the scene appear more cheerful and the town seem more lively and "prosperous."

The two photographs present such contrasting moods that the reader might think that Cambridge has changed greatly over the last thirty years. Actually, though, Cambridge has enjoyed great economic stability, judging from the size of its population, not only over the last thirty years but also over the last half-century. Its current population of 15,146 is only a few hundred more than that of 1950, and is close to the 13,104 recorded in the 1920 census.

Not surprisingly, then, Cambridge continues to "display much of the atmosphere of the smaller American town," as Stewart observed. A lack of formality is suggested in Stewart's picture by "the two sitters on the curbstone" and in the recent photo by the casual dress of the walkers on the sidewalks. The great elm is gone, likely a victim of Dutch elm disease, but the lawn and hedge which have replaced it are carefully trimmed and free of litter, probably maintained by proud gardeners. New trees have been planted in the park fronting the "old-fashioned" courthouse, just to the right of the photo view, and on the sunny noon hour that we happened by we found shoppers and business people "lounging" on benches in the shade. The sturdily braced sign facing the street has a simple message which might have appeared in many towns of 1980: "Arts and Crafts Festival in City Park."

The Civil War monument, prominent in both photographs, illustrates the patriotism that characterizes small towns in America. The trash receptacle, visible in the recent photo beside the old white stone milepost of the National Road, repeats that national pride with its conspicuous star and its red, white, and blue colors. Fire hydrants and posts supporting parking meters elsewhere along the street are similarly painted. Also not visible is a new memorial beyond the Civil War monument with nineteen names listed below the inscription: "In Memory of Persons from Guernsey County, Ohio . . . who died during the Vietnam Conflict."

Even though Cambridge has features which reflect what Stewart describes as "the enduring tradition of the small town," it also displays characteristics of a larger urban settlement. The whitewash on the walls of the first building on the left has been stripped off, revealing the stylish natural color of brick. A more elegant facade of pillars and windows, a "modern store front," has been added to the corner portion of that building, which houses the drug store. The "baby skyscraper" which Stewart identified at the extreme right side of his photo has been demolished, and when we took the 1980 picture workmen were busy with noisy construction machinery excavating a depression for a basement and foundation of a modern bank building. More wires clutter the space over the street; modern streetlights arch over the roadway; and newly planted trees grace the street and sidewalk. Automobiles continue to "clatter through the town," and traffic has become sufficiently congested to prompt the city leaders to provide off-street parking, as evidenced by the sign immediately to the left of the war memorial which announces, "Downtown Park and Shop."

S-BRIDGE

S-Bridge

About nine miles west of Cambridge, Stewart climbed a slope adjacent to Fox Creek. He looked westward over both an S-shaped bridge on the old National Road and the straight modern alignment of U.S. 40 disappearing into the haze. The highway had recently been expanded, but only two lanes were open to traffic. Stewart found it necessary to explain that the "rather startling appearance of the large truck proceeding down the wrong lane is therefore fallacious."

Although colorful legends developed to explain the origin of the S-bridges, the shape of the structure was a rational engineering solution to the problem of building a bridge over a creek in the early 1800s. The shortest distance across a stream is a straight line at a right angle to the channel, and such an alignment also simplifies the cutting and laying of stones of an arch bridge. If a road intersects a stream channel at an oblique angle, and if a right-angle crossing of the channel is desired, the S-shape is a necessary result. The key, then, is that only the middle portion of the structure is actually arching over the creek, and the end portions are merely walls holding fill to permit approaches to the true stream crossing. The slow rate of traffic movement in the early nineteenth century made the shape practical, although, as Stewart noted, it represents "almost the perfect visualization of a modern highway engineer's bad dream."

In 1980, the bank on which Stewart had stood was heavily overgrown with saplings of elm and sassafras, and a carpet of poison ivy. We were thus forced to stand lower on the slope to get an unobstructed photograph. The highway has the same alignment and design it had in Stewart's day. It may actually carry less traffic than it did thirty years earlier, because U.S. 40 is paralleled by Interstate 70, immediately south of the forested hill in the background.

The old bridge remains sturdy and unblemished, even though it is now 152 years old. Stewart had predicted that the growth of woody plants along the wall of the bridge was the initial stage of its decay: ". . . saplings are beginning to sprout along the parapet and will soon crack it." He apparently did not foresee the use and maintenance of the bridge and its environs as a park with the associated pruning and outright removal of plants that might threaten the bridge or visitors' views of it. This recreational function is revealed also by the picnic table, mowed grass, and planted ornamental trees and shrubs beyond the far end of the bridge. The long-cantilevered streetlight, at first glance seemingly extended in the wrong direction, functions to illuminate the little park rather than the modern highway.

The bridge serves users other than the recreational motorist who may stop to admire the historic structure or to eat a lunch. As in Stewart's picture, tire tracks on the bridge indicate that it still carries occasional traffic. We saw a farmer drive his pickup down the lane in the foreground from the hill to the right and use the bridge as a shortcut to reach the highway. Shortly thereafter, three small children accompanied by an older man we assumed to be their grandfather walked down to play beside and beneath the old structure. One of the little girls was carefully collecting a small bouquet of Queen Anne's lace from plants growing along the edge of the roadway, against the bordering wall of the bridge.

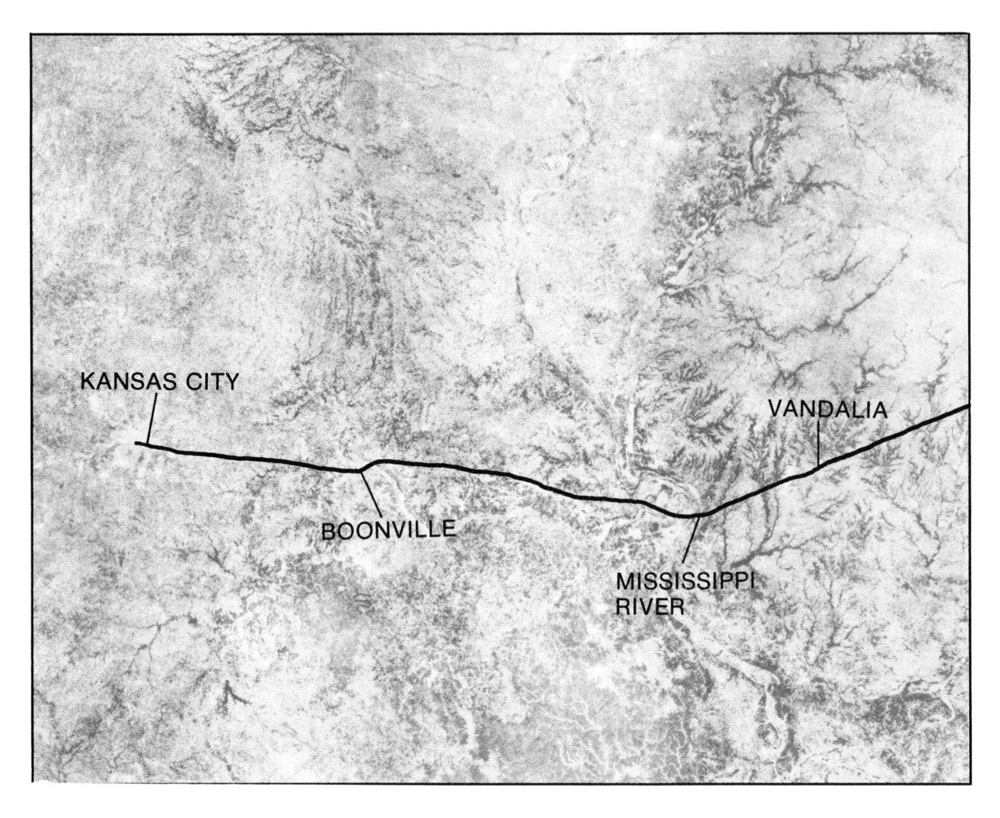

COLUMBUS TO KANSAS CITY: MIDDLE WEST

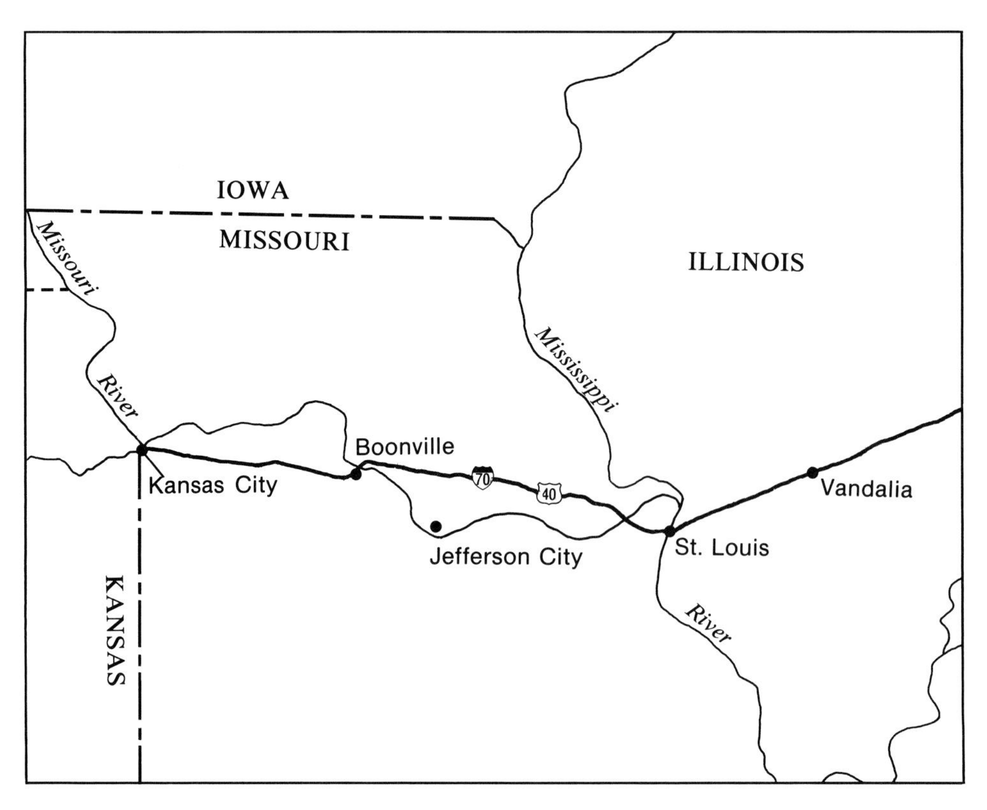

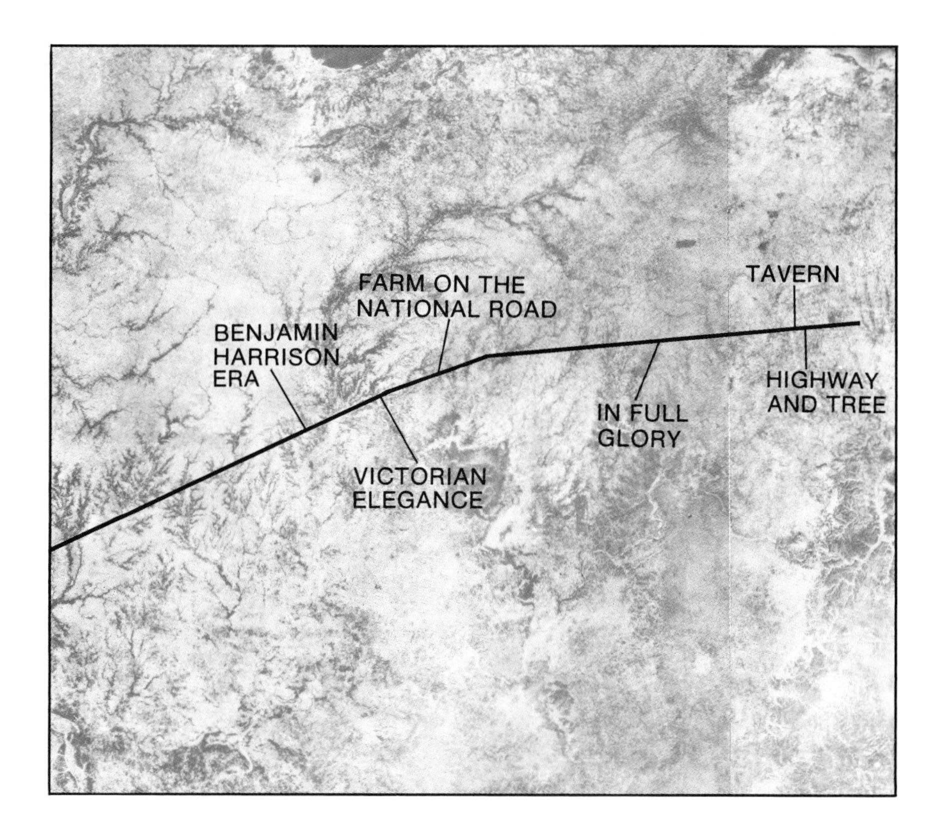

FARM ON THE NATIONAL ROAD
TAVERN
BENJAMIN HARRISON ERA
HIGHWAY AND TREE
IN FULL GLORY
VICTORIAN ELEGANCE

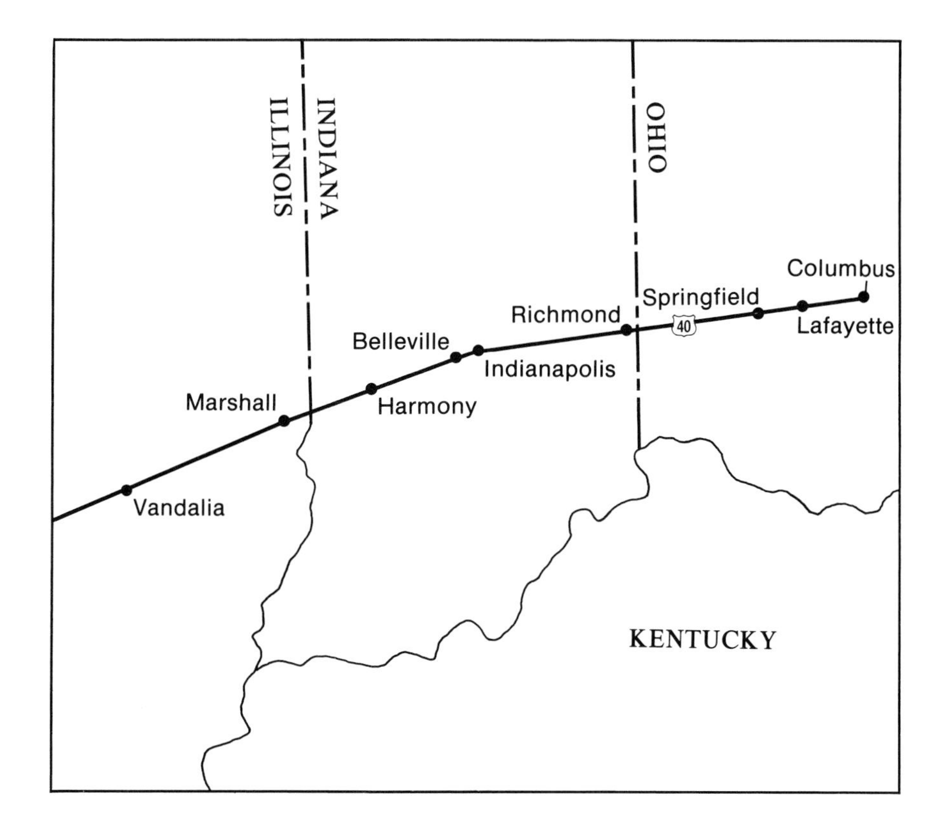

ILLINOIS
INDIANA
OHIO
Columbus
Springfield
Richmond
Lafayette
40
Belleville
Indianapolis
Marshall
Harmony
Vandalia
KENTUCKY

Columbus to Kansas City: Middle West

IN CENTRAL OHIO, U.S. 40 westbound leaves the dissected upland of the Appalachian Plateau. From near Columbus to just west of Kansas City, a distance of over 700 miles, the highway crosses the great corn-growing region of the United States. This segment of highway is characterized by flat to gently rolling country, large fields of corn and soybeans, innumerable small agricultural towns, and hazy summer skies sometimes shattered by exciting thunderstorms. Across Ohio, Indiana, and Illinois, Stewart commented that U.S. 40 "offers little of scenic interest . . . , [presenting] nothing that is conventionally picturesque." In eastern Missouri, the highway passes through rolling terrain, but overall "when all is said, nothing stands out. The ordinary tourist, having made the drive, would probably be hard put to tell anything, except that he had twice crossed the Missouri River." Still, for the observer who admires the greens of growing things highlighted by the soft browns of newly plowed fields, placid cattle grazing in neat pastures, and large old farmhouses set beneath shade trees and behind expanses of manicured lawns, this region may seem beautiful. "What the stranger might call monotony," the geographer J. B. Jackson has said of the Middle West, "the native, more flatteringly, was inclined to see as a kind of wholesome simplicity and lack of artifice."[23]

U.S. 40 on this stretch is secondary to Interstate 70. From Columbus to near St. Louis, the U.S. 40 of Stewart's time remains so designated today, with the interstate freeway running parallel, usually at a considerable distance. The U.S. route serves the local or regional traffic needs, running through the towns rather than around them. Across most of Missouri, the alignment of the U.S. 40 of 1950 has been occupied by the freeway, and the highway carries the designations for both the U.S. and the interstate routes.

Fortunately for the preservation of its unusually historic character, the eastern portion of the midwestern segment of U.S. 40 has not been modernized into the interstate highway. From Wheeling, West Virginia, to Vandalia, Illinois, U.S. 40 is on the alignment of the extension of the National Road. Surveyed in the 1820s (recall that the National Road reached Wheeling in 1818), the road was not constructed beyond the Appalachian Plateau until 1830. It was intended to extend all the way to

Jefferson City, Missouri, but actual construction became intermittent beyond Springfield, Ohio. Growing interest in railroads, and rivalries between towns that wanted to be located on the route, both contributed to the slow progress. As a result, the right-of-way, laid out in essentially a straight line, was established and some clearing of trees was completed only as far as Vandalia, Illinois, which is recognized as the western terminus for the National Road.

Stewart commented about the importance of this segment of the National Road:

> Petering out toward the west, the extension of the National Road never attained the greatness of the original. It served the central parts of Ohio and Indiana, rather than the nation as a whole. . . . [But the road] must have had an effect in detaching the people of these states from their early southern connections and allying them with the East instead of with slave-power of the lower Mississippi Valley. As we must always remember, the most important freight that a road carries may be neither household goods, nor livestock, nor munitions of war—but ideas! . . . [In addition,] in time and in place, it attained greatness. Moreover, it is safe for the future. From Columbus to Indianapolis and from Indianapolis to Vandalia the original route, whether called U.S. 40 or something else, will furnish the main-traveled road until someone disproves the geometrical proposition that a straight line is the shortest distance between two points.

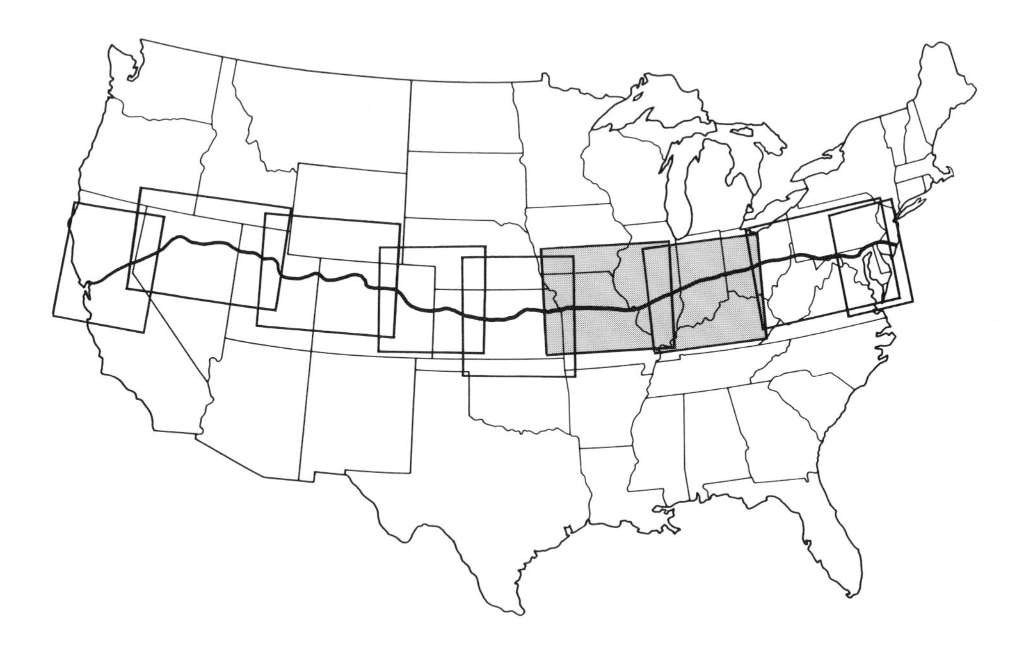

HIGHWAY AND TREE

Highway and Tree

West of central Ohio, U.S. 40 leaves the wooded and more rugged terrain of the Appalachian Highlands, and enters the Central Lowlands, a geographic province that has been characterized as having "a great deal of sameness and plainness."[24] From Columbus to St. Louis the highway crosses landscapes typified by gently rolling or flat topography and large farms of corn and soybeans. This is the heart of the Corn Belt. It is the core of the prosperous Midwest.

About fifteen miles west of downtown Columbus, Stewart stopped to photograph a large sycamore protected behind a neat rock wall and arched out over the pavement. The modern highway was on the original alignment of the National Road, and the Ohio Department of Highways had widened the roadway to four lanes, but apparently by crowding them into the original eighty-foot right-of-way. Stewart remarked that the resulting narrow four-foot dividing median was "inadequate . . . for high-speed traffic," and that the encroachment of mailboxes onto the shoulder was likewise a safety hazard. He did find, however, many aspects of the highway design to be good: a wide and graveled shoulder, a "large and well-constructed drainage ditch," and a grassy and gentle backslope rising from the ditch. "Most interesting and pleasing is the careful protection of the fine sycamore tree. With the re-grading of the highway, this tree would have been sacrificed except for its protection by a stone wall, in itself an attractive bit of masonry."

In 1980, not only was the sycamore still alive but it also was the same general size and shape. Pruning by highway crews is probably responsible for the almost rectangular gap in the tree's foliage visible above the road in both photographs. The otherwise full, rounded form suggests that the tree, which is growing on an open, sunny site, is healthy.

Several highway features noted by Stewart have changed in ways which reflect the neglect of U.S. 40 since construction of the parallel route of Interstate 70. Two continuous lines of weeds have been permitted to prosper in the cracks which separate the lanes of traffic from the dividing median. Although the shoulder is now paved, small trees and nonwoody plants thickly cover the backslope and completely hide the drainage ditch. Even the stone wall protecting the tree is masked by a fronting growth of plants.

Two other features at this site represent changes since Stewart's day. The home on the rise above the highway to the left has a broad expanse of mowed lawn sweeping down to the highway. Such large areas of manicured grass characterize farmsteads today in the upper Midwest, and their increased popularity since 1950 reflects not only attitudes toward gardens but also the advent of power mowers. On the opposite side of the highway, the field behind the sycamore supported corn three decades ago, but was planted to soybeans in 1980. Soybeans have only recently become a major crop plant in the United States, but today they characterize much of the agricultural landscape in the southern Middle West and in the Mississippi River Valley.

SINCE 1837
THE Historic
RED BRICK
TAVERN
DINING ROOM

TAVERN

A HOUSE OF HOSPITALITY
RED BRICK
TAVERN
OPEN 11:00

Tavern

PAST AND PRESENT, travelers on the road need to eat. Before the days of the fast-food franchises, taverns such as this one served that need. It is located in the small village of Lafayette, about twenty miles west of Columbus, which is beyond the horizon in the distance. Built in 1837 and said to be the second oldest Ohio tavern, the Red Brick Tavern enjoyed early prosperity during the heyday of travel on the National Road. Six presidents dined here, including five who served between 1830 and 1850, a period during which the National Road across Ohio was still the preeminent means of travel. As with many other such business establishments along the National Road, however, this tavern closed when the developing railroad system took commerce away from horse-drawn wagons and carts. During this period, a Zanesville, Ohio, audience was told that "the glory of the old road has departed. . . . Its life was short; its work is done; its greatness is of the past."[25]

When Stewart passed through Lafayette, the Red Brick Tavern had been reopened and probably looked more tidy and clean than in its earlier period of prosperity. Stewart was not exaggerating in his description of the scene: "Architecturally beautiful, ivy-grown, shaded by its magnificent sycamore, neatly fenced in white—the Red Brick Tavern suggests the charm of the past." Both the heavy traffic on the highway and the smooth well-maintained appearance of the road surface itself in the 1950 picture suggest the resurgence in road travel associated with the automobile, a resurgence which undoubtedly led to the reopening of the tavern.

By 1980, the scene was far less elegant. The brick walls and chimneys remain sturdy, but the vines which once covered the front and part of the side are dead and gone. The great shading sycamore is likewise no more. The porch lights flanking either side of the front door are rather casually and unattractively held to supporting brackets by twisted lengths of bright copper wire. The sidewalk and curb are weedy, and the formerly neat fence has been broken, as has one of the decorative lights which border the gate to the lawn area. A carelessly driven auto may have caused this damage, and such an accident also would explain why the sign which announces NO PARKING ANY TIME is twisted to face the wrong direction.

The decline in the prosperity of the Red Brick Tavern may involve the siphoning off of traffic from U.S. 40 to Interstate 70, which here is less than a mile to the north. There are other indications of the reduced importance of the highway since Stewart's day. Traffic is light; in fact, we had no trouble standing for many minutes at a time in the middle of the road to take the photograph. The pavement, which seems to have been relatively newly laid in 1950, now is worn and patched, with the expansion joints of the underlying concrete lined with tar. The "jiggle bars" which formerly discouraged vehicle movement across the narrow dividing median are gone, perhaps because lighter traffic made their function obsolete. Their former presence, however, is outlined on the road surface. The overhead flashing lights that warn motorists on U.S. 40 to slow down at the intersection were probably installed during a busier period before completion of the interstate.

But the view retains attractive characteristics. The tavern building remains "architecturally beautiful," and even the elimination of the rear side door was done while carefully preserving the continuity of the brick wall. The lawn is mowed and free of weeds. A bench, painted white to match other wooden features of the building and grounds, has been added beside the walk. The old National Road mileage pillar, just beyond the broken portion of the fence, is still present and offers a feature of historic interest. Business is apparently still good, judging from the eighteen autos parked for lunch on the far side of the building, but the fact that all eighteen have Ohio license plates suggests that local people rather than tourists are now being served.

Our stop included a meal, as apparently did Stewart's, and we enjoyed not only the food but also the old-fashioned interior of the tavern. In our leisurely wanderings on the grounds after eating, we discovered English sparrows nesting behind the lintel over the front door, and swifts darting around in the sky overhead. Wire screens over the chimney openings, visible in our photograph, apparently keep these birds from utilizing the chimneys of the tavern.

IN FULL GLORY

In Full Glory

The loss of agricultural land to expanding urban development is what many Americans feel underlies the most striking recent landscape change in their country. In our cross section of the United States, we have seen it near Middletown, Maryland, and will see it later at Denver, Colorado. Proposals for such zoning changes may meet with hearty approval from city fathers, proud to have their community grow, but with some resistance from small farmers, saddened at the prospect of still more rows of corn plowed under to make way for a cluster of roadside businesses. Yet in comparing photos illustrating such urban development, both developer and preservationist may temporarily lay aside their judgments of good or bad to share observations of the physical details of change.

The view here is west toward Richmond, Indiana, with the oldest part of town well beyond the low hill in the distance. Since 1950, the city has spread eastward over the hill, across the lowland, and beyond to the junction of U.S. 40 and Interstate 70. The location at the outskirts of town is suggested by the presence of automobile dealers, who typically need large areas of inexpensive land to park vehicles. Lots for the auto dealer on the left and the adjacent bowling alley are cut down to the level of the highway. The parking areas for both businesses are set well back from the road, apparently beyond the white right-of-way post seen in Stewart's photo. That this is an edge-of-town view is also suggested by the motel sign, with its room rates partly hidden behind the second streetlight pole. At the crest of the hill but beyond our view are the franchise restaurants and fast-food outlets which characterize recently urbanized American roads, as well as still more auto dealers, one of whom is flying the large American flag.

The highway of U.S. 40 has been modified to accommodate the new roadside businesses and the increased traffic they have generated. The concrete pavement of 1950 has been resurfaced with asphalt, although the expansion joints in the older roadway still show. Right-turn lanes allow vehicles to pull out of the flow of traffic and slow down before turning. Wooden poles supporting streetlights are close together, and must illuminate the roadway at night unusually well. The steel guardrail in the foreground protects the increased number of motorists from the supports of the railroad overpass, from which the photo was taken.

The truck weighing station is conspicuously absent. Cross-country truck traffic now follows Interstate 70, which bypasses Richmond, and thus the station is no longer needed here. But its presence in the earlier photo also reflects outmoded highway standards. Today the placement of structures such as a weighing station within the dividing median would be considered poor design. The confusing pattern of lanes connecting the station with the highway's inside, or fast, traffic lanes suggests why modern standards require weighing stations to be on the outer, slower, side of the highway with simpler entry and exit ramps.

Otherwise, however, the general alignment and layout of the highway shown in Stewart's photo are much the same as today. Lanes remain twelve feet wide, with unobstructed shoulders and drainage ditches. Three crossovers remain sufficient for side access. This permanence supports the comment in 1950 by the State Highway Commission of Indiana that if the road were to be constructed again, "there are no improvements that would be made over the present design."

Stewart, who noted "a lamentable scattering of debris, including a long strip of paper in the foreground," would be pleased to see that the median and shoulders are here, at least at the moment, unlittered. Moreover, they still regularly receive close, neat mowing, and therefore lack even a single stem of the blue chicory or white Queen Anne's lace that we saw in such profusion along Ohio's stretch of U.S. 40. The used-car dealer on the left has added to the tidy appearance of the scene by landscaping with rocks and small shrubs. The trees along the creek have grown gracefully tall, and mingle with trees more recently planted in new businesses and residences.

FARM ON THE NATIONAL ROAD

Farm on the National Road

We had real difficulty finding the "Blue Ribbon Farm," which Stewart described as being "near Belleville, some twenty miles west of Indianapolis." After two hours of driving back and forth on either side of Belleville, looking for the farm that should have been on "a rise to the south" of the highway, we pulled into the Friendly Inn, one of the few businesses in the tiny town. Several women, drinking coffee and chatting before heading home for the evening meal, recognized Stewart's picture almost immediately. They told us the site was just west of the town limit and beyond a small new housing development, but that the farm no longer appeared as it did in the 1950 photo. Following their instructions, we were able to get our companion photograph.

Stewart described the farmhouse as "plain" but admired its "fine" proportions, the "general symmetry and the classical motifs" of the Greek Revival doorway, and the "harmonious" arrangement and size of its windows. He did recognize that the house seemed "to look backward, and showed little evidence of active living," and that its size, too large for a contemporary family, was "an anachronism." In 1980, the house was not just behind the times but was gone entirely. In its place, we found an equivalently "plain" mobile home apparently permanently set on the low hill.

In contrast to its predecessor, this house does not have a "windowless end," and it lacks "the large chimney at either end [which] bespeaks the cold weather of the Middle West." Rather, an end window allows a view to the east, and the house is heated by propane, as the tank beside the house indicates. This source of heat likely reflects a population too scattered for the installation of natural gas pipelines. The unadorned and functional style of the new home may be taken to be a modern addition to the three stages in house development Stewart identified along the National Road. The first homes were built of unmodified logs; next appeared houses of hewn and squared logs; and finally came houses of sawed lumber or bricks. The aluminum construction of the modest single-story home which today stands on a little rise above the old pike may be an example of a more recently developed fourth stage.

The remainder of the farmstead in 1980 also suggests a decline in elegance. The "well-painted fences" are gone, and have not been replaced even by barbed wire. The wire fence fronting the highway in 1950 is also absent. The low arch bridge still crosses over the creek, but no longer displays a freshly whitewashed exterior. "Excellent" cannot be used to describe, as Stewart did, the single barn that remains, which is, curiously, the older of the two standing in 1950. The look of the farmstead suggests that it may no longer function as a working farm, but may house an individual, perhaps with a small family, who commutes to a nearby town for work.

In contrast with the rest of the farm, Stewart found the house and its immediate surroundings showed "little evidence of active living." In 1950, "one of the sheltering trees [was] dying, and the others call[ed] for attention." Today, none of those trees is alive, although a sapling has grown up near the new detached garage. The four evenly spaced trees visible along the edge of the highway in 1950 have also been reduced in number.

The highway in both photographs appears unchanged. As in Stewart's day, the nearer lanes have been recently resurfaced, and the dividing median "has been allowed to grow up charmingly with late-summer flowers." This similarity in appearance, however, belies the great change in traffic over this stretch of U.S. 40. As noted earlier, Interstate 70 across Ohio carries most of the vehicles, and U.S. 40 serves the local or regional traffic. The economy of the small town of Belleville has thus suffered twice by transportation changes, first by the completion in 1850 of a railroad which moved traffic off the National Road, and more recently by the construction of the interstate which, according to a book on Ohio place names, "left Belleville stranded on the old highway to the north, greatly reducing the life-giving tourist trade."[26]

VICTORIAN ELEGANCE

Victorian Elegance

About thirty miles west of the farmstead near Belleville, Stewart stopped to photograph a fine old brick house built in 1872. The National Road had been extended across Indiana in the 1830s, and thus the original owner built his home long after the initial settlement by Europeans in this area. Its large size and elegant style, as well as its location in hilly wooded terrain, suggest that it was not part of a farm. Perhaps the first owner had accumulated wealth from mines in the great coal field which extends from about this point westward across most of southern Illinois.

Stewart called the photo "Victorian Elegance," and said the structure represented "much of mid-Victorian sophistication and comfort." But in describing the subdued features of the house, such as the "simple" struts supporting the eaves of the facade with its "admirable" proportions, Stewart noted that "there is less fussiness than one generally associates with the period." He commented that originally "fine trees stood between house and road," but that "with the widening of the highway and lowering of its level, the house has been left standing on a high treeless bank," with only a hedge for protection.

Like the farmstead, this old home has experienced an accelerated decline since Stewart's day. Today the house is abandoned, and the formerly protective trees and shrubs are beginning to engulf it, even to the extent that a locust tree blocks the front door. The eaves are rotted, hornets fly among the struts, and swifts dart in and out of the chimneys. None of the windows is broken, although each in the bottom row is tightly shuttered. The driveway rising from the highway to the right of the house is bordered by overhanging shrubs and has a locked gate upon which a sign proclaims emphatically and simply: "Absolutely No Trespassing; Anyone Caught Will Be Arrested."

We did not disobey the sign on a warm July afternoon when we stopped for our photograph. We did receive, nevertheless, two additional warnings against trespassing into the old house. The first came from a young boy who was riding by on his bicycle as we were wandering along the edge of the westbound lanes visible in the picture. He told us that the house was owned by his uncle, who formerly allowed Halloween parties in the old structure before he became concerned over possible vandalism. A few minutes after he pedaled off, the county sheriff drove up and repeated the warning. He said that the police had been frequently called to investigate alleged break-ins, and that someone apparently had used a shotgun against people who were trespassing.

Ironically, all of this happened only one mile east of a town named Harmony. This name apparently refers not to the personality of the place but rather to the pleasing sound of the word, as explained in *Indiana Place Names:* "There is no reason to be assigned for the naming of this town and the postoffice . . . other that that of euphony and suggestiveness."[27]

BENJAMIN HARRISON ERA

Benjamin Harrison Era

Marshall, Illinois, like Cambridge, Ohio, developed as a town on the National Road. Situated at the junction of two railroads and serving as the seat of government for Clark County, Marshall has maintained itself as an agricultural and administrative center without great periods of boom or bust. Its rate of population growth has been steady but slow; the 1980 population of 3,500 is not much greater than the 1900 population of 2,077. Probably in part because of the absence of great demographic or economic fluctuations, the row of downtown buildings in the photograph has not been demolished, and it continues to house businesses that serve the everyday needs of the citizens of the town and farms.

Stewart spoke glowingly of the buildings: ". . . the main business-district is in its way an architectural gem, and might well be preserved as a historical monument. . . . In a hundred years, if [these buildings] should be preserved so long, people may be comparing them with the Grande Place in Brussels or some of the crescents in Bath."

In general, the solid, well-painted structures have retained their dignity through the first third of the century of Stewart's concern. Yet, there are some signs that they may eventually suffer from neglect. None of the glass panes are broken, but two in the third building from the corner have been replaced with wood or metal. On the window panes of the building to the right, peeling paint allows the rays of the sun to penetrate the glass and reach the rooms within. Some of the window frames show exposed wood and thus may be rotting. Bric-a-brac no longer are stacked behind the quaint lettering "Antiques" above Grabenheimer's; like most of the other upstairs floors, the area appears to be either vacant or used only for storage.

The architectural changes are minor, but they do detract from the uniformity of style and period that Stewart so admired. The skyline is now broken not by the large ornate sign indicating the 1889 erection of the corner building but by a single television antenna. Gone is the row of cloth awnings of an earlier era. Only one awning, made of aluminum, remains. Beyond the drug store still operated by the Blankenships, a new contemporary front of wooden shingles has been added to a clothing store. Two doors farther down the street, the closely set pillars on the lower facade of a variety store are partly covered by new siding. The most drastic change on the block is the absence of the final building in the row, which was demolished and replaced by a new structure, set back from the street and therefore not visible, which houses a coin-operated laundry.

Other changes do not threaten the historic personalities of the buildings. Stewart remarked that ". . . during the summer this southward-facing row of buildings must be hideous with heat," but today air conditioning may make that concern outdated for the occupied sections of the buildings, even without the shade-casting awnings. Only two small air conditioners, in buildings toward the end of the block, are visible. The streetlights and traffic signals are modern, although utility wires serve the buildings from the back. Stewart developed interest in, and appreciation for, utility wires, but their absence on this street was probably important to his admiration of the buildings. He questioned the necessity for parking meters, and today they are gone, although the supporting posts remain. The sidewalk at the corner has inclined ramps to permit bicyclists to cross the curb onto the steet more easily, as the cyclist in the 1980 photograph is about to do. The street has been resurfaced with asphalt, thereby covering the brick exposed in the earlier photo.

Like Stewart, we arrrived in town during the noon hour. The quiet of the time was interrupted by occasional vehicles, overalled farmers hungry for lunch, and shoppers wandering from store to store. To the left of the photo view, three older men were sitting on a bench in front of the county courthouse. We wondered if residents still gathered on the lawn of the surrounding park to hear music from the small bandstand under the sheltering trees. The sky was dotted by fluffy cumulus clouds, recently formed in the cool air following a front that had ended a long siege of hot and humid weather. It was one of the few days when we had a more interesting sky than did Stewart.

VANDALIA

Vandalia

In about 1960, the Center for the Study of Democratic Institutions wanted "to find out what was going on in a small American town—a town that was not a satellite of a big city, a town not dominated by a single industry, a town with a strong historical heritage."[28] The Center chose Vandalia.

It was the "historical heritage" which prompted Stewart to take his photograph of this scene. Vandalia was the end of the final extension of the National Road. Originally envisioned as extending at least as far as the Mississippi River, seventy miles west of Vandalia, the National Road was instead halted here because of local rivalries for routes and the growing importance of railroads. In the photographed view, the road came from the east down Gallatin Street from the right and ended at the corner in the foreground, where the statue of the "Madonna of the Trail" overlooks the spot. Although in Stewart's day U.S. 40 still followed this route, today it adheres to the alignment of the National Road into Vandalia only as far as the corner one block east of the one in the photo. From there, U.S. 40 turns to intersect Interstate 70 north of town.

The second major historical feature which prompted Stewart to take this photograph is the old state capitol building. Vandalia was the seat of the state government of Illinois from 1819 to 1839, a time when young Abraham Lincoln was serving as a state legislator. This particular old building, however, was not built until 1836, and it was modified during the twenty-year period after the state government moved to Springfield. In Stewart's time, as today, the capitol was a museum designed "to re-create the interior visual surfaces of the 1836 period, [while] retaining . . . the exterior appearance of 1858." In our 1980 visit, we found that the building resembled the structure in Stewart's photograph, although a few of the plants in the grounds surrounding it had changed. The elm to the left of the building had grown and was casting heavier shade and blocking our view of the capitol. The tree to the right of the flagpole in Stewart's photo had been removed and replaced by another tree, a sweet gum, itself now grown handsomely large.

We encountered a gardener and his helper digging up an attractive, apparently healthy shrub in front of the pillars of the portico. "A forsythia doesn't fit into the 1836 period," he explained. Similarly, he said, the master plan for the park calls for removal of juniper trees and other shrubs which were not characteristic of ornamental planting in the mid-1800s. He also indicated that the hedge and low concrete wall fronting the sidewalk were to be eliminated, perhaps to be replaced by a wooden board fence and walkway. As at Fort Necessity, concern for accuracy in historic reconstruction was the guideline, although, again as at that earlier historic site, we suspect that the manicured lawn will remain.

The recent photograph reveals much of the atmosphere of a small town. The absence of signals at the corner indicates light traffic. A modern "Madonna of the Trail" "strides forward on her way" with stroller and leashed dog and crosses the street outside the crosswalk without apparent concern. Stewart implied that the parking meters in his photo seemed inconsistent with the "relaxed atmosphere of the small town in summer," and today they are gone, although the posts remain as they did in Marshall. As the president of the Vandalia Chamber of Commerce commented during the 1960s study of this typical small American community, "Things around here don't change very much from one year to another."[29]

MISSISSIPPI RIVER

Mississippi River

From a vantage point on the Illinois side of the Mississippi River, the skyline of St. Louis, Missouri, is framed beneath the intricate cantilever spans of the old Municipal Bridge. Because the river was much higher in June of 1980 than when Stewart took his photograph, we were forced to take our photo from farther up the bank.

The history of St. Louis is a particularly vivid instance of the early vigor, later decline, and recent uncertain rejuvenation of many American cities. Paul Nagel writes that the citizens of St. Louis in the 1870s foresaw their city as "the logical new leader of America and perhaps the globe." But such optimism faded quickly, and by the turn of the century St. Louis, neither grown into a "leader of America" nor interested in being a leader of its more immediate hinterland, had become "infamous for its tenements, filth, violence, and corruption."[30] By Stewart's time, after more than fifty years of continued decline, St. Louis had reached "a kind of nadir," and Stewart's photograph seems to capture the mood of that unhappy condition. The dark bridge hangs over the low, gray buildings of town. A deserted waterfront borders on the featureless waters of the river. Extending over everything else is an overcast sky, dull and seemingly still, accentuating the somber feeling of the scene.

Over the following two decades, St. Louis enjoyed a revival of spirit and activity, and the more recent photograph illustrates some of the results. Busch Stadium, noticeable for its round walls with a crown of arched fluting that partly hides the city hall, represents the effort to bring people back into the downtown area. Several other new office and residential buildings break the skyline, including the PET building, which completely blocks from view the Southwestern Bell Telephone Company building that was so conspicuous in Stewart's photo. The new bridge that carries a constant flow of traffic following both U.S. 40 and Interstate 70 over the Mississippi is simple in design, having nothing of the "lace-like" pattern which Stewart admired on the Municipal Bridge. Barely visible at the extreme right side of the recent photograph is one end of the Gateway Arch, a memorial to western expansion in the United States. Near its base is a landscaped area of grass and trees, and a parking lot full of cars. The sky is overcast, as it was in Stewart's view, but the scene seems full of light and presents a positive mood.

What we saw on two visits to the waterfront in St. Louis during 1980 came close to persuading us that the city was recovering from its "mortal illness." This area, with its arch, boat rides, riverboat businesses, and other amusements, was crowded with tourists. Several old blocks of brick buildings on the waterfront district north of the arch were being sandblasted clean and rebuilt internally for offices and businesses. Freight trains and river barges came and went carrying the products of commerce. But the appearance of vitality is deceiving, because the vigor and optimism that characterized St. Louis after 1950 was brief. By the 1970s a critic could again say that the city was "literally a crisis in human dependence and poverty."[31] The 1980 census gave dramatic credence to his assessment: the city of St. Louis lost nearly one-third of its population during the 1970s, a rate of loss far exceeding the decline in other major American cities. Urban renewal projects may be successful in promoting the city as a nice place to visit but not necessarily as a desirable place to live.

In fact, we found that even for visitors, if they wander from the beaten tourist tracks, St. Louis may create a small "crisis in human dependence" and a new appreciation for the squalor of the impoverished areas. When our Volkswagen broke down on the freeway, we traveled on foot and by buses, loaded down with cameras, film, and notebooks, across the city to the airport to rent a car while ours was being repaired. With dusk falling, we waited at bus stops or wandered along streets in seedy neighborhoods, expecting the worst. As the streetlights went on, we quickly made the decision to pay the $5.00 cab fare for the remaining five-minute ride to the airport.

The side of the river from which the photo was taken, on the other hand, has an entirely different atmosphere from the highly urbanized side of St. Louis. Woodland, field, and pasture extend well back from the Mississippi, probably undeveloped because the area lies within a levee which protects East St. Louis farther east from flood waters. Birds sang from tall cottonwoods and elderberry thickets, and the water lapped on the sandy shore. On the day Stewart stopped for his picture, "an aged Negro was quietly fishing," and in 1980 we encountered an elderly black man and a young boy, probably his grandson, walking leisurely toward the water, fishing poles resting on their shoulders.

BOONVILLE

BOONVILLE

MIDWAY ACROSS the state of Missouri, Stewart stopped in the town of Boonville and photographed the Thespian Hall. Built in the 1850s, the structure was "the oldest surviving theatre building west of the Alleghenies." Stewart admired not only its origins as a cultural center but also its role as a military hospital and prison during the Civil War. He seemed particularly intrigued with its use in 1950 as a movie theater, "presenting to contemporary citizens the dominating folk-art of the mid-twentieth century." But he decried the "hideous and unnecessary modern sign," which announced the building's new name, the *Lyric*.

In 1980, the Boonville Community Theater group was using the Thespian Hall, which had regained its original name. The "hideous" sign had been replaced by a more modest one attached to the wall over the door. Other changes similarly suggested a refurbishment of the building. The white paint which had covered the bricks of the pillars had been stripped off, exposing the natural color and patterns of the masonry. Decorative shutters had been added to the side windows. The brick pavement which in 1950 had extended from the bottom step to the edge of the concrete sidewalk was torn up, but the bricks in 1980 were neatly stacked against the wall of the building down the side street visible in the photo view, presumably stored for later replacement. Containers with Wandering Jew and geraniums were attractively hanging from the streetlight pole in front of the hall.

The building beyond the Thespian Hall was also admired by Stewart as "an almost equally interesting example of mid-nineteenth century Americana." In 1980, the structure was being gutted, although the main exterior walls were still standing and presumably would remain. The three features which Stewart so admired, however—the pointed arch of the porch, the roof-pinnacles, and the top-heavy chimney—were gone. It seems likely that offices will eventually occupy the space, and not a commercial retail business such as the bakery thrift store which is housed in a more modern building farther down the street. The tall elm trees, which seem not even to be present in Stewart's photo, will also likely be retained.

Some other features of the view have changed little or not at all. The stone marker commemorating a battle of the Civil War is still present, as is the fire hydrant, although the latter has been turned so that an outlet points toward the sidewalk. The streetlight standard is more modern, and it no longer supports the shield of U.S. 40, even though the highway still follows this main street through town. In the foreground of the more recent photo, the original brick pavement of the road is visible where the covering of asphalt has been broken and worn away.

KANSAS CITY

Kansas City

The view northward from the top of the Liberty Memorial Tower presents a panorama of downtown Kansas City, Missouri. For Stewart, the scene was "the portrait of the typical American city as traversed by U.S. 40." Similarly, many of the changes depicted in the two photographs are common to urban centers of the United States over the last thirty years.

The massive Union Railroad Station remains, although its functions have, in part, changed. In 1950, Stewart commented that "the parking spaces and lines of taxis in front of the station may be taken as symbolic of the present-day dependence of railroads upon motors." By 1980, the decline of the railroad and rise of the automobile was even more complete, as ironically indicated by the emptiness of the parking lot and the complete lack of taxis. The long coverings over the passenger walkways beside the tracks are visible in Stewart's photograph to the right of the station but are gone in the 1980 photo, also suggesting the decline in rail traffic. Like railroad stations elsewhere, this one now houses other businesses, although it continues to serve the occasional railroad traveler. Among the new businesses are two restaurants with bills-of-fare typical of such railroad station eateries: steaks and seafood.

The growth in the importance of the automobile is suggested by the reconstruction of the Main Street overpass which crosses the railroad tracks. New concrete support pillars have been completed in the area immediately east of the station, and other pillars are being constructed farther out over the tracks. Stewart had commented that the streetcars on Main Street in 1950 were "worthy of note since streetcars in an American city are already beginning to carry something of an antique or quaint value." By 1980 they were gone, and the reconstruction of the overpass will certainly not provide for them.

New buildings diversify the skyline of the downtown. Most of the new structures house commercial services, offices, and residences, often together in the same building. These are functions of the new rectangular structure east of the railroad station, which, unfortunately but perhaps characteristically of some American cities, has been built in what was formerly the open space of a park.

The "secondary business district," lying between the downtown and the railroad station, appears little changed. The red brick buildings, which Stewart accurately described as "architecturally utilitarian . . . flat-topped, almost cubical, boxes of commerce and industry," remain mostly functional and well maintained. We noticed some boarded windows within the district, but only a few buildings seemed abandoned or run-down.

Stewart decried the presence of the billboards in the vacant lot at the lower right corner of his photograph. In 1980 they were not only gone but actually replaced by lawn, trees, and walkways of a park. This parkland is the edge of a great residential-hotel complex which occupies the same hillslope that helps to give the Liberty Memorial Tower such a fine view of Kansas City.

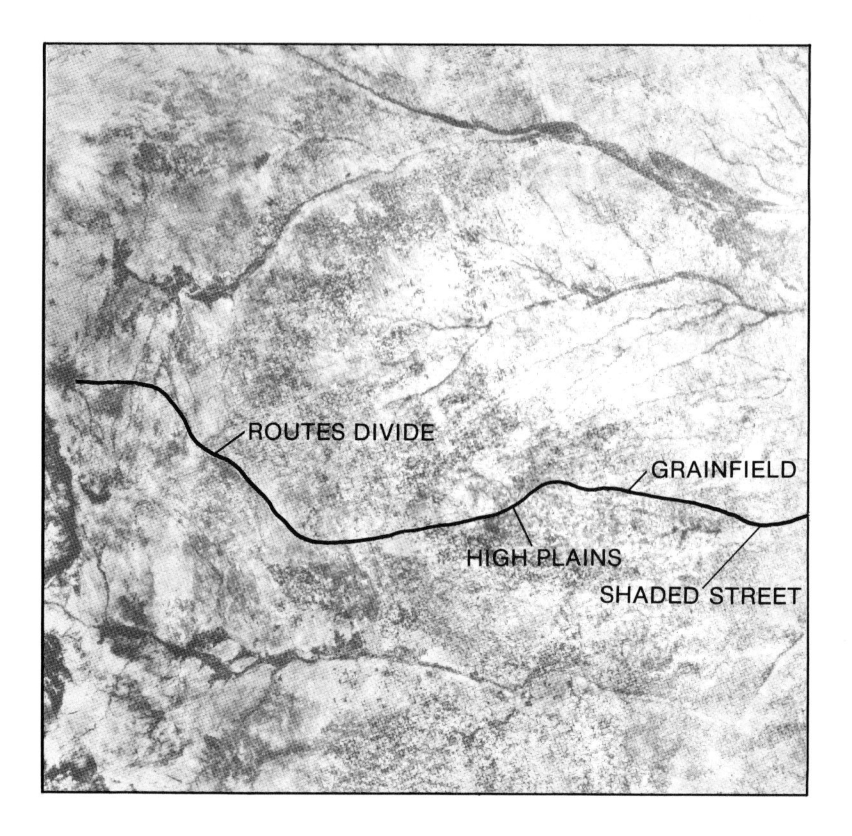

KANSAS CITY TO DENVER: EAST BECOMING WEST

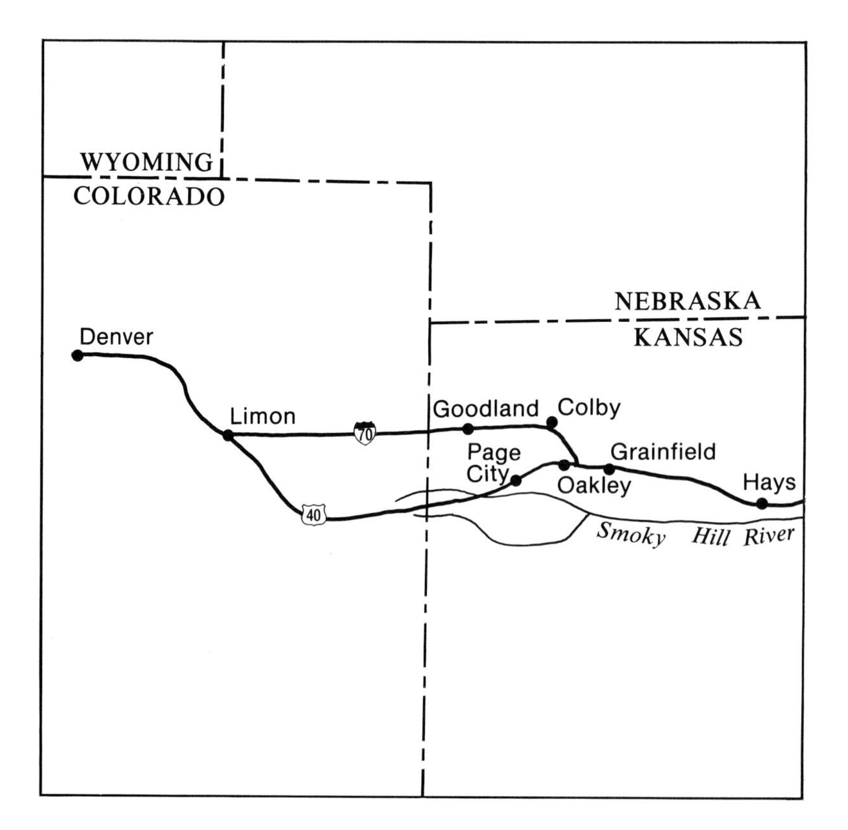

ROADSIDE PARK
KANSAS CORN
HAYFIELD
PLAINS BORDER

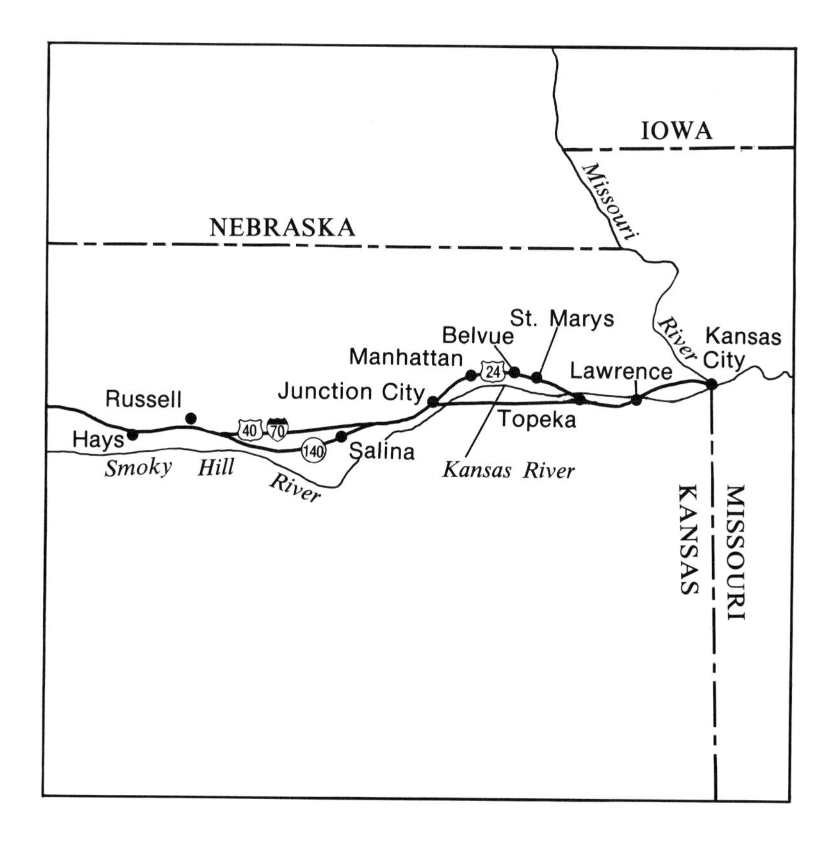
IOWA
Missouri
NEBRASKA
St. Marys
Belvue
River
Kansas City
Manhattan
24
Lawrence
Junction City
Russell
Topeka
40
70
Hays
140
Salina
Smoky Hill
River
Kansas River
KANSAS
MISSOURI

Kansas City to Denver: East Becoming West

IN 1950, U.S. 40 was one of three main U.S. highways which headed west from Kansas City. U.S. 24 followed a northerly alignment, passing through mostly small and little-known towns, whereas U.S. 50 dipped far to the south, toward Hutchinson and Dodge City. U.S. 40, adhering to the drainages of the Kansas and Smoky Hill rivers, maintained a course more directly west, and passed through most of the largest settlements of northern Kansas. It was the main highway to Denver.

Today, Interstate 70, continuing its role as the modern replacement for U.S. 40 across the country's midsection, keeps mostly to the historic route followed by its predecessor. The freeway often deviates slightly from the U.S. 40 of Stewart's day, passing, for example, to the north of Abilene and to the south of Russell, but only in two stretches is the relocation major. Between Topeka and Junction City, U.S. 40 in 1950 kept close to, and north of, the Kansas River by following a broad arc through St. Mary's and Manhattan. Similarly, west of Salina U.S. 40 swung to the south in a gracefully curving route, adhering closely to the Smoky Hill River. The interstate, with its function of speeding travelers along as efficiently as possible, has cut off both bends and thus follows a more direct east–west line. Beyond Oakley, in far western Kansas, in contrast, the freeway leaves the U.S. 40 of Stewart's day as well as of today and jogs to the north, where it intersects and then follows the route of U.S. 24. U.S. 40 is left as a rather minor highway which falls to the south, crosses into Colorado, and then swings northward to meet the interstate again at Limon. Reunited, the two route numbers follow the old alignment into Denver.

The highways have undergone some changes in the last thirty years, but the landscape through which they pass continues to present "a drama more striking, perhaps, than any to be seen elsewhere along the whole cross section—the gradual drying up of the country, westward." From the fields of corn, small woodlots, and closely spaced towns in the moist climate along the Kansas-Missouri border, the highway passes into the expanses of wheat, great grasslands, and far-flung settlements which characterize the more arid realm of western Kansas and eastern Colorado. The air seems drier with each mile, and, with little to break the long vistas, the arching blue canopy overhead often dominates the landscape. Ground features may become so subordinate, in fact, that, as Willa Cather said of the western prairies and plains, the earth sometimes becomes "the floor of the sky."[32]

The sky and land produce a dramatic country, particularly toward the west. With better economic incentive, there is more wheat grown than in Stewart's time, and more irrigation of other crops, including corn. More traffic crowds the highway, but the towns, even with their larger and more numerous grain elevators, remain small and quiet. These settlements, first visible from the freeway as a line of dark trees in this otherwise treeless landscape, are usually touched only on their fringes by the highway. Most travelers rush on across the plains without slowing down or turning off, and see nothing but interchanges and a few roadside businesses. But when the Front Range becomes visible on the horizon as the highway rises toward Denver, the traveler's eyes are often drawn back to the golden or reddish stubble of harvested wheat, the gleaming white of a nearby cluster of grain elevators, or a flock of pronghorn grazing on the short grasses. "On the right kind of day—and there are many of them—you can watch the great white clouds piling up, and massing, and ever-changing. Then you inhale the great openness of the country and feel the spirit expand, and—as you keep your foot well down on the throttle—you can see, hour by hour, the East gradually shifting over to be the West."

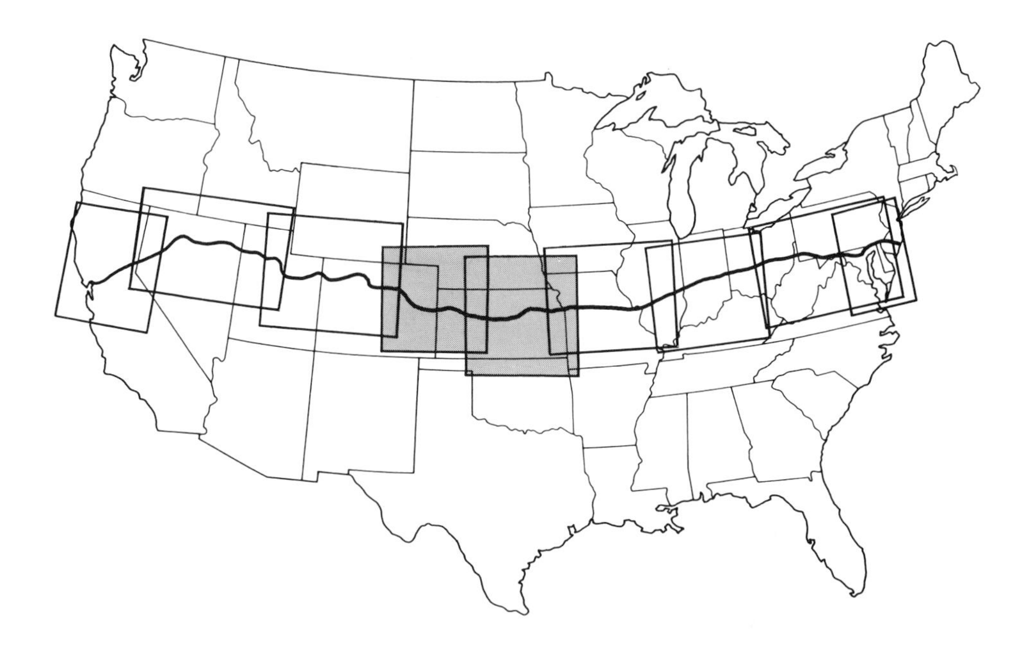

HAYFIELD

Hayfield

Locating Stewart's hayfield, described simply as somewhere "between Lawrence and Topeka," was not easy on a hot and windy summer afternoon. Changes in land use and lack of conspicuous landscape features required us to drive back and forth over several miles of road which Stewart had said was on "the level plain a few miles back from the trench cut by the Kansas River to the north." But we finally did find the field, then planted in corn. Not knowing which of several nearby homes might house the owner, and reasoning that our trespassing would be for only a few minutes, we climbed a fence and walked out along a property line to set up for the photograph. The planted corn in the field and the volunteer shrubs and weeds at its edge were too tall, however, to permit exact duplication of Stewart's view, so we scouted for a companion scene that allowed a view of the features which Stewart had recorded. This delay increased our trespass time, and as we were about to climb the fence to leave, we noticed a young farmer hurriedly driving his tractor toward us from a modern home across the highway. Although at first understandably angry, he soon became interested in the old photograph and the people portrayed in it as we apologized and explained our activities. Eventually we were invited to the home of his parents, Mr. and Mrs. Raymond Ice, who lived at the end of the gravel drive, where we enjoyed iced tea and friendly conversation about the fields in Stewart's photograph.

The two men tending the hay bailer in Stewart's photo, we learned, were young Raymond, squatting at the rear, and his father-in-law, driving the machine. The family had moved to their ridge-top farm in 1946, and thus had been there for only a few years when Stewart passed by. They planted hay in this former pasture in their first years, but eventually switched to corn. The corn production from the upland site had not been consistently good, however, and the dry summer weather of 1980 had already almost precluded a crop when we stopped in mid-July.

The sparsity of woody growth in Stewart's day had not changed much by 1980. Although small trees now line the north side of the highway where none grew in 1950, the scattered trees and shrubs south of the highway that were present thirty years ago are now gone. Stewart had interpreted the absence of trees as reflecting a dry climate: "The region here is far enough west so that tree-growth is becoming scanty, except in the stream-valleys and moister lands. The look of the country to the right of the highway suggests that it was never wooded." Our conversation with the Ices suggested, however, that conditions are hardly too dry for woody plants. The "never-wooded" pasture south of the highway had been cleared of shrubs only a year or two before Stewart took his photograph. Moreover, the Ices indicated that they frequently cut back the "scrubby trees," as Stewart described them, which grow up along the fence lines. Evidently a landscape does not always reveal to the eye the forces that have acted upon it.

Other features in the photographs also show both persistence and change over the last thirty years. The telephone lines strung on five cross arms in Stewart's photo have been replaced by three stout cables, but many of the poles are identical. The highway remains a "pleasant little backwater . . . winding intimately across the farming country," although the new interstate highway is only a few hundred feet to the left of the photograph view. The land along this stretch of U.S. 40 continues to be devoted mostly to the growing of corn and hay, but new residences, like the Ice home across the pavement to the right, have sprung up like scattered trees along the highway.

As we left the Ice family, Raymond was heading out to join his son in the nearby field. Their job that afternoon was to continue the harvest of alfalfa hay, just as it had been for a much younger Raymond Ice and his father-in-law on another summer afternoon thirty years before.

KANSAS CORN

Kansas Corn

One-fifth of the distance across Kansas, Stewart stopped west of the little town of Belvue, walked up the highway overpass by which U.S. 40 crosses the railroad, and photographed a scene of Kansas corn. He commented about the strong popular association between that crop and that state: "The words of the song, 'As corny as Kansas in August,' are well demonstrated here. . . . Except for a single field, everything that is cultivated is in corn."

Actually, Kansas was not in 1950, nor is it today, a major corn-producing state. As was true in Stewart's day, it does not even rank among the ten leading states in corn acreage. Corn production in Kansas in both 1950 and 1980 was close to that of Georgia, which is hardly thought of as a major corn state. Admittedly, far-eastern Kansas is the western edge of the great Corn Belt of the Midwest, through which our trip has taken us since we emerged from the Appalachian Plateau in central Ohio. But Ohio, Indiana, Illinois (the number-one producer), and even Missouri all devote much larger acreages to corn than does Kansas. The view in the photographs is near the western boundary of the westernmost county in Kansas which supports unirrigated corn, and even in this county, Pottawatomie, only 14 percent of the harvested cropland is planted to that crop. Most of Kansas is, quite simply, too dry for corn.

Economics have also influenced crop plantings in this scene. In Stewart's day, Pottawatomie County had more than twice as much acreage in corn as today; since then, other crops, particularly sorghums, have gained in importance. Appropriately, then, most of the field area south of the highway in the recent photo is planted to sorghum. The other crop which has gained greatly in Pottawatomie County since Stewart's time is soybeans, which is probably the crop newly planted in the field in the foreground.

When song lyricists write about agriculture in Kansas, they should talk about wheat. In 1980, as in 1950, Kansas was the country's leading wheat state. From about this point west to the Rocky Mountains, wheat becomes as conspicuous a landscape feature as corn is between central Ohio and eastern Kansas. The view in the more recent picture, had it been photographed a month earlier, would have illustrated a stand of wheat ready for harvest in the field recently planted to soybeans.

Besides the changes of crops, another obvious difference between the two photographs is the large grain elevator in Belvue in the 1980 view. Stewart had noted the "huge silo" on the right side of the older photo, and even though it is still present today, it is dwarfed by the newer, much larger grain-storage structure. These towering grain elevators characterize most of the towns in the wheat-growing region farther west, and they will become a distinctive feature for the next several hundred miles of U.S. 40.

The grain elevator is conveniently located beside the railroad, by which grains eventually move to market. The continued use of the railroad is suggested by the cleared right-of-way immediately adjacent to the tracks.

A heavy growth of woody plants along the edges of the right-of-way is yet one more indication that the climate will support trees. That woody plants are otherwise generally absent in this landscape, except in the town of Belvue, is owing to the gentle terrain and productive soil which have encouraged the large fields of wheat, sorghum, and corn which typify eastern Kansas. The vigorous growth of woody plants along the railroad suggests how rapidly shrubs would move in if the cultivated fields were abandoned.

ROADSIDE PARK

Roadside Park

Less than one-quarter mile west of the overpass, a narrow strip of land between the highway and the railroad continues to function as a roadside park. The gently curving access road has been unnecessarily widened and its gravel surface replaced by asphalt, but overall the park's facilities remain modest. Most traffic uses Interstate 70, more than ten miles south of here, and only local traffic follows U.S. 40 on its broad arching course along the Kansas River.

The general appearance of the park has changed dramatically over the last thirty years, reflecting the demise of its largest trees, that "first necessity of a roadside park," as Stewart suggested. The massive cottonwood trees which he admired for their size and cool shade are gone. Only the outlines of their rotted trunks remain in the sparse grass. Farther along the park road, the trees that had been small in Stewart's day have grown noticeably taller, but none even approaches the former dominating presence of the cottonwoods. To the left of the highway, the several trees which Stewart identified as recently planted are, like the cottonwoods, gone, perhaps victims of a road-widening project. The park could hardly be said to support "a fine grove," as Stewart predicted, although the growth has produced the not-unattractive look of a thicket.

Many physical features in the park have survived better than the cottonwoods. The rest rooms, "modestly set among a planting of trees" in 1950, are now completely hidden by the woody growth. The pump remains functional, and refreshed these 1980 travelers with a good stream of fine cool water. The oil-drums still serve as litter barrels, but today they have been modernized with domed covers with swinging doors, plastic liner bags, and signs which proclaim "Pitch In."

Several other facilities in the park have been added or changed since 1950. Just beyond the pump, a new shed that houses maintenance equipment and supplies has been built in the same style as the rest rooms. A distinctly contemporary touch is the sturdy display board with its sloping shingled roof. When we stopped by in 1980 it contained a sun-faded and water-stained road map of Kansas. Even though not visible in the recent photograph, picnic tables are scattered through the park. Not of the old-fashioned style illustrated in Stewart's picture, they are modern steel-framed structures with wooden tops and seats. Several are set on concrete slabs, either to eliminate muddy ground underfoot or to provide a firm anchorage. One such concrete mass can be seen at the left edge of the recent photo, but its table has been dragged farther to the left to some shade provided by a small tree.

This roadside park, or roadside rest as it would be called today, is rather primitive by modern standards. It has no inside plumbing, flush toilets, disposal facilities for recreational vehicles, watered and mowed lawns, structures to shelter picnic tables from sun or rain, or lighting to aid night-time travelers. Still, whether modern or primitive, whether located along a four-lane freeway or a two-lane highway, whether in 1980 or 1950, such roadside parks continue to function as Stewart described: "Here the tourists—particularly in family groups—can stop to eat a picnic lunch at one of the tables, let the dog find a tree, and give the children a chance to play about and stretch their legs, while the car cools off in the shade."

PLAINS BORDER

Usually we were able to locate the general locales of Stewart's photos rather easily, although finding the specific spot where Stewart had stood was frequently difficult. Determining even the general area of this scene west of Salina, however, was a frustrating puzzle. We were not able to solve it during our first stop there in June of 1980. Success came on a hot, dry, windy day later that summer, after driving back and forth for several hours along a ten-mile stretch of highway in the empty rangelands of the Kearney Hills.

Most of the difficulty stemmed from our expectation of finding the treeless landscape which Stewart had emphasized: "Rainfall . . . is here about twenty-seven inches annually. The upland country in this area has always been treeless, and if buffalo were substituted for cattle, the distant view would be just about what the earliest explorers saw." Stewart's conclusion that general conditions in the area had changed little since a century or two before was probably true, and this view is the first on our east-to-west trip along U.S. 40 about which such a statement can be made. Nonetheless, the growth of trees over the last thirty years makes the foreground of this particular landscape look considerably different from that in Stewart's photo. Trees are actually common from this eastern edge of the Kearney Hills eastward, but west of this point, while U.S. 40 rises and falls gently through the broken topography, trees are restricted mostly to ravine bottoms. Trees in the photograph, mostly willows and an occasional cottonwood, one of which is the tallest form to the right, are growing in a shallow ravine which extends across the view, draining from left to right. That swale was present in Stewart's day, but it is not easily identifiable in his photo, perhaps because clouds obscured the shadows which would have highlighted it. The ravine may be more moist today than thirty years earlier, however, because sometime during that period a low earth embankment was built across the drainage and now impounds a shallow pond, presumably for livestock, immediately to the left of the photo view. The seepage from that reservoir may be enough to encourage the growth of woody plants, although the abundance of trees just to the east suggests that the climate is probably sufficiently moist to permit their growth.

The dominant land use within the Kearney Hills continues to be livestock grazing. No Herefords were standing in the hot sun of the day when we took our photograph, but many were in the shade of large cottonwoods beside the shallow pond. The area is too far west and too dry for unirrigated corn, but it is not too arid for wheat. Perhaps the dissected topography discourages the growing of such an agricultural crop in the Kearney Hills. Whatever the reason, the appearance of rangeland here foreshadows the broad expanses of western landscapes still hundreds of miles away.

The highway has been obviously resurfaced recently, and thus lacks even a center line. Most traffic follows the freeway designated as both Interstate 70 and U.S. 40 ten miles to the north, but local traffic is surprisingly heavy at this point, probably because of the nearby state park and reservoir on the Smoky Hill River. The new fence, with its characteristic metal white-tipped posts, is thus needed to keep cattle from straying onto the pavement. The decrease in the status of the roadway, which is now Kansas State Route 140, would not explain the loss of the "magnificent pole-line," as Stewart described it, but the line's replacement, undoubtedly by underground cables, adds further to the feeling of isolation of the scene.

A grove of trees in a broad expanse of grassland, as portrayed in the recent photograph, offers some people a feeling of security and permanence in the wide-open vistas of such landscapes, according to novelist Wallace Stegner. The trees' protection may be particularly welcome when bitter winter winds sweep unhindered across the empty spaces. Blowing strongly on our day in the Kearney Hills, however, were the warm winds of July. Of such prairie winds in summer, Stegner writes: "Across [the] empty miles pours the pushing and shouldering wind, a thing you tighten into as a trout tightens into fast water. It is a grassy, clean, exciting wind, with the smell of distance in it. . . . It turns over . . . every pale primrose, even the ground-hugging grass. It blows yellow-headed blackbirds . . . and ruffles the short tails of meadowlarks on fence posts. In collaboration with light, it makes lovely and changeful what might otherwise be characterless."[33]

SHADED STREET

Shaded Street

Like Dodge City, one hundred miles to the south, the name of Hays, Kansas, conjures up images of the old American West. Near the eastern edge of the Great Plains, Hays is the largest town between Salina and Denver, and thus serves as a commercial center for those characteristic products of the western plains, wheat and cattle.

In this view eastward along the main street of Hays, Stewart focused his discussion on the shade trees. The climate, he explained, is too dry for easy establishment of trees, but "if certain varieties are given a start, by being watered through the first few years, they establish themselves and continue to grow well." Stewart observed that east from Hays residential areas of most towns have treelined streets, but farther west such streets are typically "almost treeless." He further suggested that commercial streets, regardless of location, are almost always without trees.

Since Stewart's time, the elms in the foreground have died and have not been replaced. Other trees farther down the block survive and still provide shade, although the feel of a continuous canopy of foliage seems lost. Other residential streets in Hays continue to support fine stands of street trees, particularly elms.

The apparent expansion of commercial businesses into this former residential area may contribute to the changed look of the neighborhood. The businesses on the north side of the street are small grocery stores, gasoline outlets, and sandwich shops. They may have been established in this part of town in order to serve students at Fort Hays State University, which is about two blocks farther west.

The influence of the university is suggested by other characteristics along this street. Although some of the dwellings seem to be homes for established families, as evidenced by their good maintenance, neatly mowed lawns, and trimmed hedges, other houses are more reminiscent of student housing, with peeling paint, old furniture on the porches, and less-manicured gardens. The amount of litter was surprising, given the street's general neat appearance; we found beer bottles and soda cans in the gutter. The house for sale was not vacant when we stopped for the photograph, but the grass growing in the badly cracked walkway and crowding over the edge of the sidewalk is in sharp contract to the tidiness displayed in Stewart's photo.

Even so, except for the loss of some trees and the expansion of commercial businesses, the street appears much as it did in 1950. The brick pavement is the only example of that paving material surviving from the U.S. 40 of Stewart's day. Utility poles and wires, absent except for streetlights in 1980, probably were absent too in 1950, although the thicker growth of trees in the earlier view makes that a matter of speculation. The utility wires, as in many towns of America, run along the rear property lines and thus are hidden from this vantage point.

We photographed the scene early on a Sunday morning in June. The long shadows suggest the time of day, as do the empty sidewalk and quiet street. The only continuous activity was the loud but pleasant chattering of western kingbirds, fly-catching over the street from perches in the trees. The presence of that bird, as much as the dry air or the lack of extensive wooded terrain, told us that we were at the edge of the American West.

GRAINFIELD

GRAINFIELD

GRAINFIELD

THE GEOGRAPHER J. B. Jackson identified three features which contribute to the distinctive character of the wheat country of western Kansas: the grid pattern of the township and range survey system, the large and labor-efficient farm, and the dominating importance of the railroad. Two pairs of photographs of the small town of Grainfield, both looking east, illustrate the persistence of these features.

The large rectangular fields and the long straight lines of the railroad, highway, and utility lines illustrate the widespread, lasting effect of the grid system. The great size of the farms is revealed not only by the large areas of individual fields, but also by the general emptiness of the landscape. Except for the dark spots created by clumps of sheltering trees, one would scarcely suspect the presence of widely scattered houses. The critical importance of the railroad is suggested by the line of buildings strung out beside the tracks; these structures serve Grainfield's primary economic activity, the growing of wheat.

The town itself conforms so well to Jackson's description of western Kansas wheat towns that it seems as if he were describing Grainfield specifically:

> The tracks come straight out of the East: there are the elevators, the depot . . . the freight yards, the warehouses; and as near to the tracks as they can get are the wholesale builders supply, the feed dealers, the dealers in agricultural machinery. At right angles with the tracks is Main Street, with a hotel on the corner near the depot, and the post office not far away. The grain elevators are the most important buildings, rising far above the fringe of Chinese elms.[34]

The most visible changes in Grainfield over the past thirty years are simply variations on these essential themes. Perhaps most conspicuous is the great increase in storage capacity for grain, noticeable in both of the recent photos. The tall elevators have been expanded by the addition of new rounded columns. Lower, squat storage bins have been built, and structures for the mixing of different feed grains, identifiable by the long slanting arms rising from lower storage facilities, have been erected. From these latter structures, combinations of grains, mostly sorghum, are trucked out to area farms for the feeding of hogs and cattle, as perhaps the truck on the highway in the recent photo is doing. The addition to the elevator immediately to the left of the road sign in the ground-level view prevented our standing exactly where Stewart did. Instead, we had to walk out to the edge of the new structure, and thus got a photograph from a point a little farther east.

At the time of our visit to Grainfield in mid-July, the harvest had just been completed, and the storage facilities were bulging with wheat and sorghum. Wheat was still being added to the more easterly elevator from a mountain of grain piled on the ground and visible as a light-colored mass just beyond the rail cars at the east end of the last building. Wheat was apparently also being "railed out," as suggested by the presence of the rail cars on the siding. The manager of Gove County Co-op elevator said that the harvest had been so good that he could have twice filled his elevator.

The old railroad station building visible in Stewart's photo is no longer present. We were told that it had been moved by the local Lion's Club to a park "where the girl scouts camp," but that within a night or two it was burned down by vandals.

In the more recent photograph, the freeway of Interstate 70 and U.S. 40 is faintly visible close to the horizon to the right. From the freeway, Grainfield appears as other towns of western Kansas appear from afar, a line of dark green trees overtopped by skyscraping grain elevators.

High Plains

Only forty miles west of Grainfield and less than fifty miles east of the Colorado state line, U.S. 40 passes through the tiny town of Page City. Here U.S. 40 has left Interstate 70, which swings to the north, and for nearly two hundred miles the routes are separate. They merge again at Limon, Colorado, from which they run together into Denver.

Both views look west. The unobstructed view over the high plains in the second photo pair is from a window at the top of the elevator seen so prominently in the first photographs. Stewart took his picture of the grain elevator from the metal shed with the covered rectangular opening beside the siding, visible at the bottom of the 1980 photo, but we were told that access to his vantage point would not be possible. Thus, we substituted a photograph from the top of still another tall elevator which may not have been present in Stewart's day.

Page City is little more than this series of grain elevators beside the railroad and a few homes north of the tracks. The highway swings away from the rails at the edge of town, thereby making room for the elevators, and then returns to an alignment closer to the tracks after passing Page City. The traveler is hardly aware that a settlement exists.

The High Plains as seen from the grain elevators of Page City, even more than from those of Grainfield, portray the character of the region. Truman Capote captured the personality of the plains environment in his description of the town of Holcomb, Kansas, nearly a hundred miles due south of Page City: "The village of Holcomb stands on the high wheat plains of western Kansas, a lonesome area that other Kansans call 'out there.' . . . the countryside, with its hard blue skies and desert-clear air, has an atmosphere that is rather more Far Western than Middle West. . . . The land is flat, and the views are awesomely extensive; horses, herds of cattle, a white cluster of grain elevators rising as gracefully as Greek temples are visible long before a traveler reaches them."[35] Stewart was less descriptive but more emotive in his simple statement: "In sheer grandeur and massive scenic effect the most magnificent mountains scarcely exceed the High Plains."

The general landscape appears little-changed since Stewart's time: vast fields of wheat, a few farmhouses identified by clumps of dark trees, and distant towns marked by longer lines of trees and the white masses of grain elevators. One of the few differences apparent in the recent view of the area west of Page City is a patch of burned grass beside the highway. Also, the two siding tracks present in Stewart's day have been reduced to one.

As in Grainfield and Belvue, the storage facilities for grain have been greatly expanded. Additional units have been added to the tall elevator in the photograph, and although not visible, lower storage bins have been built on its western side. Within the work area of the elevators, many more sheds and garages are apparent, as are cylindrical tanks for fuels. There are signs of some activity: a tank truck waiting beside the garage, a loaded grain truck heading for the elevator to discharge its bounty, and at least four rail cars sitting on the siding. But the busiest time of harvest is over.

The increase in facilities at Page City, as at Grainfield, indicates a thriving production of wheat on the High Plains. Stewart had warned that "agriculturally the country may be called marginal . . . in dry years it rapidly becomes a dust-bowl." For the time being, however, the full elevators and the piles of wheat on the ground make the problems of drought seem far away.

HIGH PLAINS

HIGH PLAINS

ROUTES DIVIDE

Routes Divide

The view is toward the southeast at the eastern edge of the little town of Limon, Colorado. U.S. 40, after crossing the low rise on the horizon, gradually falls down a long gentle slope to join Interstate 70 at the overpass in the foreground of the recent photo. The freeway divided from U.S. 40 slightly west of Grainfield, Kansas, and has been following a more northerly route along with U.S. 24.

The object of Stewart's primary interest in the scene was the competition for tourists between businesses along the two numbered U.S. routes existent in 1950, as revealed by the promotional billboards. "U.S. 24 truthfully declares that it offers the shortest route to Kansas City, [although] the distance saved is negligible. . . . The proponents of U.S. 40 advertise even more vaguely 'Save Time.'" Despite the twenty-mile difference, U.S. 40 remained the main route at that time, passing through "the more important towns."

In 1980, roadside businesses must still want the travelers' dollars, but no signs urge tourists or truckers to use one route or the other. The greater carrying capacity, better maintenance, and more direct routing of Interstate 70 make it the usual choice of auto-, truck-, and busdrivers, most of whom want to get across the plains as fast as possible. A glance at the 1981 *Rand McNally Road Atlas* removes any doubt about the predominance of the interstate. Over the two hundred miles of separation of the two highways, the largest town along U.S. 40, Oakley, has a population of 2,327, and it is only two miles from the point at which the routes divide. Adjacent to the interstate, the two largest settlements are Goodland, with 5,510 inhabitants, and Colby, with 4,658, both twice the size of Oakley. The greatest distance between any two cities is twenty-two miles on U.S. 40, thirteen miles on the interstate. Only one roadside rest or wayside is identified along U.S. 40, as opposed to seven along Interstate 70. The 1981 *North American Road Atlas* published by the Gousha Company lists as "points of interest" along the old U.S. route only Fort Wallace Memorial Museum, but notes for the freeway Sod Town, Kuska Museum, Flagler Reservoir, Tower Museum, and an airport. Except as a connection with various roads heading south, U.S. 40's main appeal over this section may be to those few who appreciate in Aldo Leopold's apt words, "a blank space on a map."[36]

Stewart found the scene marred by the "disreputable clutter of signs . . . even though not a house is in sight on the whole landscape." Today's signs are more conservative in number, size, and style, and there is still no building to interrupt the wide expanse of grassland. The level span of the horizon seems complemented, rather than disturbed, by the parallel lines of the interstate and its overpass.

Stewart also commented about the flora and fauna of the vicinity. Antelope were fairly common in the short grasslands, sometimes even grazing close to the highway's fringe of weeds. We saw some antelope several miles southeast of here, but more in evidence along this fairly peaceful road were lark buntings, whose black-and-white flutter accentuated the yellow pea flowers.

Although relatively few vehicles appear in the two photographs, the addition of lighted onramps, a cement median, and a widened, partially paved shoulder suggests that busier times are now the rule. A line of power poles marching across the empty terrain further reminds us of a world faster paced than in Stewart's day.

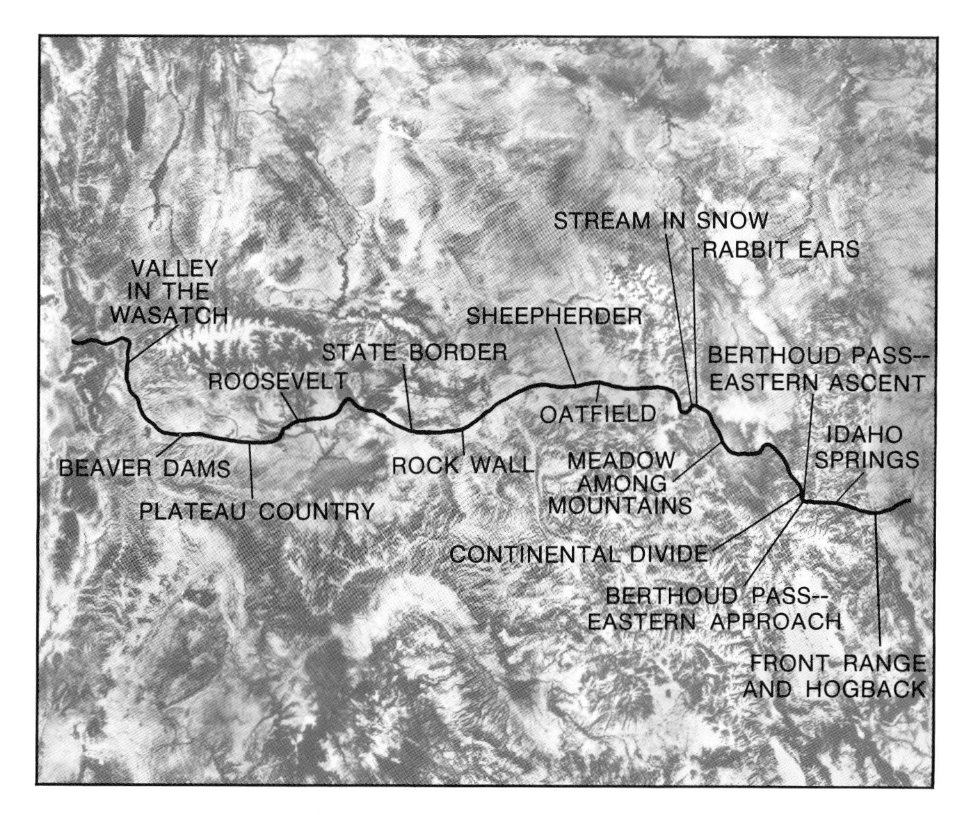

DENVER TO SALT LAKE CITY: ROCKY MOUNTAINS

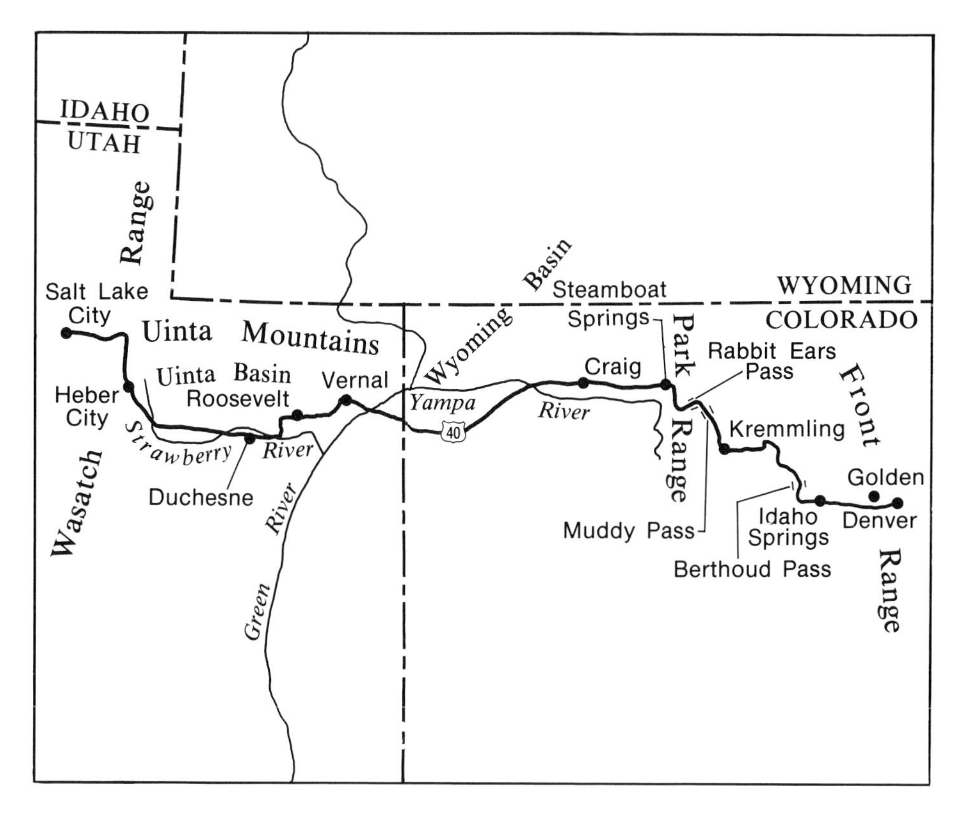

Denver to Salt Lake City: Rocky Mountains

FOR MOST OF THE 502 miles between Denver and Salt Lake City, U.S. 40 of today shares several characteristics with the highway of 1950. It is the major road of the regions through which it passes. It retains its exclusive designation as a U.S. highway, without a dual interstate identity. It is mostly a two-lane road, widening through a few towns and on some of the long steep grades. No other stretch of U.S. 40 today has these features over so long a distance.

The highway's changelessness reflects an isolation caused by the routing of the interstate freeways to far-distant alignments. Interstate 80, the main westbound freeway north of U.S. 40, crosses the Rocky Mountain region by following U.S. Highway 30 through the gentle topography of the Wyoming Basin in southern Wyoming; it then crosses Salt Lake City and eventually terminates at San Francisco. Interstate 70, with which U.S. 40 is associated westward from Wheeling, West Virginia, angles slightly toward the south after leaving Denver, thereby carrying traffic bound for Las Vegas and the urban areas of southern California. U.S. 40 is left in between, not only serving the towns and small cities of northwestern Colorado and northeastern Utah, but also providing the shortest, if perhaps not the fastest, link between Denver and Salt Lake City.

At either end of this segment of U.S. 40, however, the highway is coincident with interstate routes. It runs with Interstate 70 from the western edge of Denver for twenty-nine miles before turning sharply north to cross Berthoud Pass. Twenty-five miles east of Salt Lake City it joins Interstate 80. This particular junction now marks the official ending of U.S. 40, because from this point all the way to San Francisco the Interstate 80 emblem alone marks the route that was formerly U.S. 40; the repetition of two numbers of two different highway systems has been judged unnecessary. The ending of U.S. 40 is not noted by road signs or markers. It is modern efficiency to have one highway unceremoniously merge with another.

The terrain on this stretch of U.S. 40 is indeed "outstanding," as Stewart commented. The Colorado Rockies present a series of mountain ranges separated by high valleys; the Wyoming Basin in eastern Colorado is an interlude of rolling dissected uplands; the Uinta Basin of Utah displays an exposure of reddish rocks against the backdrop of high mountains; and the final range of the Rockies, the Wasatch, provides a grand entry into Salt Lake City. In such a varied and spectacular setting, this segment of U.S. 40 is usually regarded as the most scenic. As was true in 1950, it also "remains the least traveled and most primitive [length of] U.S. 40." We can agree with Stewart that "one should not complain about this or try to change it, but should give thanks."

Yet, not everything about this environment is simply a carry-over from Stewart's day. Signs of the recent changes, and potential future changes, in the American West are illustrated along the route. The boomtown atmosphere of Steamboat Springs, with its luxurious lodges, hillside homes, and specialty shops, is an example of the impact that affluent recreation-seekers can have on the landscape. The new coal mines and electric generating plants near Craig are reminders of the importance of the Rocky Mountain region for the production of coal, and their proximity to the irrigated farmlands along the Yampa River has created conflict between agriculture and energy development over that scarce western resource, water. Near Roosevelt, a small refinery and busy truck traffic are signs of the recent exploration for petroleum and natural gas in the intermountain West. Still, most of the landscape remains rural, lonely, and quiet, or as Stewart described it, "an empty land of far-scattered ranches and tiny towns."

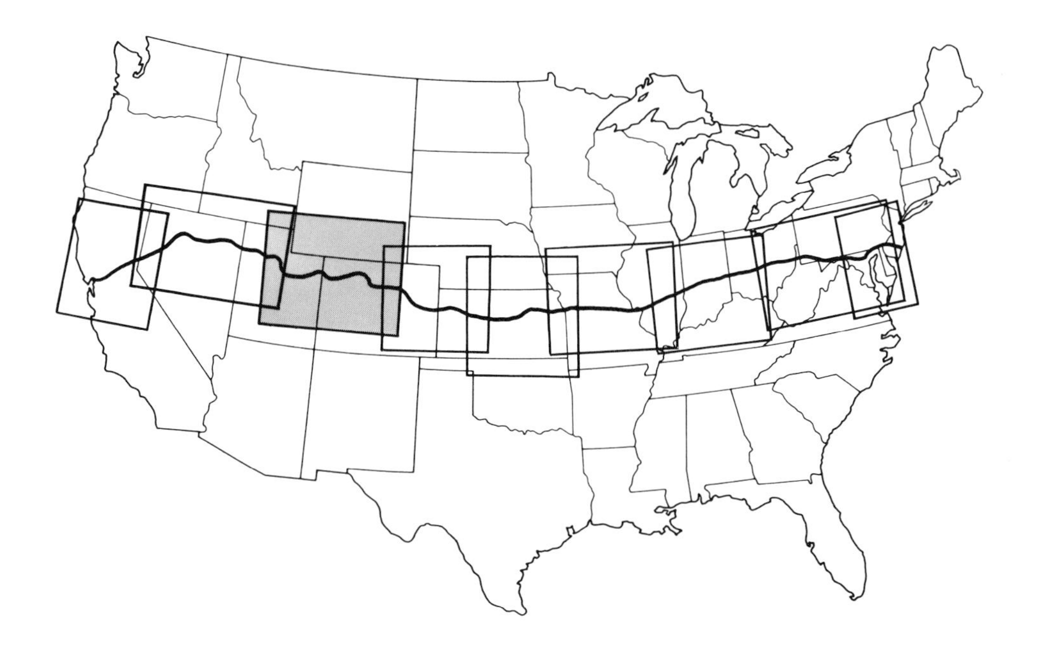

FRONT RANGE AND HOGBACK

Front Range and Hogback

After leaving downtown Denver, U.S. 40 ascends the gradual and relatively flat incline toward the Rocky Mountains. When the highway reaches the edge of the mountains, it curves southward around the northern end of the low ridge, a famous Colorado "hogback," from which the photographs were taken. Just to the left and behind the picture view, U.S. 40 turns westward again and climbs steeply up a ravine into the Front Range.

On a Sunday afternoon in 1950, Stewart observed that the highway was "dotted with cars, mostly of people coming out of Denver to spend the day in the mountains." Auto traffic is much lighter in the recent photo, not because of any changes in the vacation habits of the local population, but because Interstate 70 carries most of the cars and trucks through the ridge in an immense cut immediately south of the photo site. The stretch of U.S. 40 in the picture, then, today provides access to the westbound freeway for residents of the new houses, apartments, and mobile homes that extend across the middle distance, and for those who live in the town of Golden, hidden within the dark trees in the background in both photos.

The old highway also permits access *from* the interstate to a curiously incongruous collection of roadside businesses that reflect the persistence of rural activities in this rapidly urbanizing scene. Just east of the ridge, a garbage dump occupies an expanse of flat land. To the west, quarrying for aggregate takes advantage of exposed rocks on the first slope of the mountains, as the loaded gravel trucks on the side road illustrate. In addition, a tourist development of rustic wooden buildings located behind a low hill west of the last bend in the highway claims to offer "a unique turn-of-the-century experience in dining, shopping, and recreational pleasures." A 1900 "experience" undoubtedly responsible for many young travelers convincing their parents to detour temporarily from the interstate is a ride on a small train around a pond visible at the left side of the recent photo.

Like the surrounding flatter lands, the ridge itself has features reflecting both persistence and change over the last thirty years. The tilted rocks, exposed when the Rockies were uplifted and left as a hogback ridge because their hardness makes them resistant to erosion, appear virtually unchanged. But the trees, mostly ponderosa pines, have increased both in number and in size. Stewart indicated that the rocky slopes allowed penetration by rainwater and thus supported deep-rooted trees. While that is likely true, the expansion of the pines and the colonies of dark, low-growing brush suggests that protection from wildfires also contributes to the presence of woody plants on these slopes.

During our many visits to this photo site, we were always impressed by the contrasts in personality of the scene, an unusual blend of the rural and the urban. We enjoyed the wild character of the ridge, protected as an open-space reserve, when we walked out along its spine, keeping our eyes open for the rattlesnake that we never saw. The strong scent of pines and the melodious songs of towhees and meadowlarks often filled the hot summer air. Swallows glided overhead, and butterflies fluttered about. At the same time, we heard the background noise, mostly below us but from all sides, of humans busy at work or at play: the constant roar of the traffic on the interstate, the intermittent crashing of garbage trucks disgorging their loads, the frequent squealing of gravel trucks braking to a stop before starting up again, and the occasional high-pitched toot of a train-ride whistle.

IDAHO SPRINGS

Idaho Springs

On a hot July morning, we scrambled and slid up and down, back and forth, through tangles of brush and grass along the steep north-facing slope from which these photos were taken. Whereas this slope supports a healthy young forest of Douglas fir and ponderosa pine, most of the vegetation on the south-facing slopes across the canyon remains "only a scattered growth of pine," hardly changed in size or extent from the earlier photo. The forest is thickest on the shaded exposures, where evaporation is less severe, and these stands were probably the sources of timber originally for building homes and mines and later for supplying products shipped by rail to supplement ore shipments. Regrowth was likely as slow then as these photos indicate it is today, as even roots were grubbed out for fuel wood. The trees in town, many of which are drought-resistant cottonwoods, also appear little changed, except for those removed, along with the houses that they sheltered, to make room for the interstate. The unvegetated, eroded slopes of the tailings from the Argo Mine add dramatically to the effect of aridity.

The town itself also seems to have changed little in overall appearance through the past thirty years, as perhaps it has over most of its existence. It still "straggles" along the canyon, as Stewart observed; a recent history of Idaho Springs claims that residents joke that their town is "three blocks wide and three miles long."[37] The white-painted, wooden bungalow-style houses described as tasteful architecture in the 1871 *Colorado Gazetteer* (as quoted by Stewart in the 1950s) are still prevalent in 1980. Many of them belong to families of the original builders. The Argo Mine has remained the predominating structure since early gold mining days, and today 200 mines surround the area, producing uranium, tungsten, zinc, molybdenum, and lead, as well as gold. But, as Stewart noted, the town is now primarily a resort rather than a mining community. The mill and its associated Clear Creek Mining Museum are listed in the 1981 American Automobile Association Tour Book as the major points of interest, featuring self-guiding tours, gold panning, and western gunfights. A large billboard invites the traveler to "pan for gold."

The accessibility and comparative economic diversity of Idaho Springs may explain why it survived while many other early mining towns around it became ghost towns. Since early mining days, inhabitants had recognized the appeal of their area to tourists, and children often hawked gold and silver specimens to travelers arriving on the excursion trains. A "mammoth frame building with hot and cold shower baths, parlors and dressing rooms elaborately furnished" was constructed to attract tourists to the hot springs in a side canyon about a half mile west of this photo view.[38] As the mountain road systems were improved and expanded, and automobile traffic became common over Loveland and Berthoud Pass, Idaho Springs became a resting place for travelers who wanted to avoid trips at night or during bad weather over these sometimes windy, avalanche-prone early roads.

"Progress" in transportation has been responsible for major changes in the urban landscape of Idaho Springs. In 1941, the Colorado Southern Railroad discontinued service because of declining ore, lumber, and passenger needs. The automobile had risen in importance, and the demand for improved mountain roads resulted in the train tracks being quickly "ripped up and sold for scrap iron" so that the old railroad bed and tunnels could become the foundation for the new highway.[39] Although there was some concern that tourism would decline after the completion of the interstate, which would enable travelers to continue safely through the high mountains under dark skies and stormy weather, the town has actually doubled in size since the writing of the American Tour Guide Series in 1941.

The future seems uncertain for Idaho Springs, nevertheless. From the "reduced highway" through town, the traveler today observes lines of tourist shops. Many, however, are closed on Sundays, at least one is for rent, and few seem bustling with trade, as is evident from the largely empty parking lots in the recent view. The town in 1980 seemed more tacky than tasteful to us, with mobile homes adding to its "three-mile" string of buildings. From Interstate 70, however, one's attention is caught by the many new, rambling residences perched precariously above the slope in the foreground. If, as seems likely, most of these residences are owned by affluent newcomers who formerly visited Idaho Springs as tourists, it may be that "progress" in transportation is responsible. As Idaho Springs becomes a bedroom town for Denver, the contrast between the homes above and those below is likely to become still more pronounced.

BERTHOUD PASS—EASTERN APPROACH

Berthoud Pass—
Eastern Approach

Beyond the farthest ridge visible in the photos, U.S. 40 leaves Interstate 70 and the main canyon of Clear Creek. The highway climbs steeply beside a tributary from the north before emerging into the broad and more gently sloping valley in this view. To the right, it curves in a sharp switchback and, rising toward Berthoud Pass, crosses the foreground at an elevation of about 10,000 feet.

Stewart concentrated his comments on the vegetation: "The slopes . . . are clothed almost entirely in spruce, interspersed with patches of aspen, distinguishable by their slightly lighter color." He recognized that the aspen only temporarily occupy a site, persisting "for a century or more until the conifers gradually encroach upon them and finally take over." That encroachment seems to have progressed over the last thirty years, particularly on the south-facing slope to the left of the highway and in the valley bottom. In addition, conifers seem to have invaded the brushy vegetation along the stream.

Stewart attributed the patches of aspen to fire: "The aspen-patches usually represent areas where the spruces have been wiped out by a fire. The opening immediately in front . . . is almost certainly such a scar." While frequently responsible for aspen stands, fires are not the only disturbance that favors "bushes and aspens" in this type of environment. High winds may blow over areas of conifers, and insects or diseases may kill either small patches or large expanses of shade-tolerant trees. The long strips of aspen, brush, and herbaceous plants extending down the slopes are avalanche paths; the one in the foreground, which permits the long vista beyond, begins far upslope at timberline on the ridge above and ends beside the highway in the valley bottom. Its more open character in 1980 than in 1950 probably reflects more recent avalanching.

Human activities also often create areas of disturbance that allow an increase in aspen and other sun-demanding plants. The long narrow swath of light-colored plants through the forest of dark conifers just to the left of the highway in the distance is the cleared right-of-way for a high-voltage transmission line, apparently not present in Stewart's day. If the highway itself or other cleared areas were to be abandoned, moreover, "bushes and aspens" would likely be early colonizers.

The sparsity of human developments in the scene reflects not only the public ownership of the land as national forest, but also the limitations imposed by heavy, long-lasting winter snows and short growing seasons. The motel in the valley visible in Stewart's photo has been enlarged and transformed in the 1980 view into a trailer court, not for overnight tourists but for long-term residents, probably workers for the mines up the valley to the right of the photo. The forest reveals no sign of logging, not only because spruce is not a much-desired commercial tree, but also because trees grow slowly in this landscape of short summers and steep slopes.

U.S. 40 here thus serves mostly people passing through: miners on their way to work, skiers on their way to resorts north of Berthoud Pass, and truckers and tourists bound for points farther west. Though it is not here a freeway, U.S. 40 carries considerable traffic. Actually, from the point of its separation from the interstate beyond the horizon to its junction with Interstate 80 near Salt Lake City, a distance of about 450 miles, U.S. 40 assumes the same importance as a major thoroughfare, uncompromised by a competing interstate, as it had along its entire length in Stewart's day.

BERTHOUD PASS—EASTERN ASCENT

Berthoud Pass—Eastern Ascent

No other of Stewart's pictures provides such a fine vista of mountainous terrain as this view "from a mountain spur 11,500 feet in altitude [toward] the continental divide around the headwaters of Clear Creek at an altitude of about 12,800." We parked our car at the summit of Berthoud Pass and climbed up the ridge through a forest of spruce, where we surprised an elk foraging near timberline, and then walked across the alpine slope until we found what we were convinced was the rock on which Stewart had stood. Low, early morning clouds to the east forced us to wait for more than an hour in a cold wind before some clearing allowed us to photograph the scene. The soft backlighting and lack of cumulus clouds of our morning are in contrast to the highlighted landscape and puffy clouds of Stewart's afternoon. The abundant snow in the recent July photo testifies to the late arrival of warm temperatures in these high mountains; the nearly snowless slopes of Stewart's September photo probably were to be soon dusted by the first snow flurries ending the short summer season.

From this ridge, as from the hogback earlier, the almost constant rumble of autos and trucks centered our attention on the highway below. Not everyone would agree, nor would we suggest, that all mountain slopes are improved by roads, but here we find the highway to be a visual asset. The nearly six miles of curving switchbacks are in their own way as impressive as the mountains through which they pass and add a graceful line to the landscape. Stewart did not make such an observation, but we think that he might have agreed.

The boldness of the highway is accentuated by the raw, unvegetated cuts and fills. Over the last thirty years, the road right-of-way has become invaded by trees, particularly noticeable on the highest-elevation large cut, but most of the earth that was bare in 1950 is still exposed today. Moreover, the cuts and fills in Stewart's photo seem nearly as raw as those of a newly constructed highway, even though this road was completed in 1938, years before Stewart passed by. Such lack of revegetation is in sharp contrast to the establishment of trees and shrubs on road cuts in the eastern states. It cannot be simply that the climate is too harsh for tree establishment, as indicated by the thickets that have thrived in a few spots. Rather, the unstable soil on the cuts produces slides of rock and earth, particularly during spring snowmelt and summer thunderstorms, and thus prevents woody plants from maintaining toeholds. The common practice of dumping slide material from the pavement down the fills also contributes to the continued rawness of those slopes; we observed many rocks on the pavement during our drives along this stretch of highway.

If the road cuts are often disturbed by slides, the forests through which U.S. 40 passes are less frequently disturbed by fire than earlier in this century. Protection from fires was revealed in the view of the eastern approach to Berthoud Pass by the increase in conifers at the expense of aspen. Here, the main evidence of fire control is the general thickening of the forest. In their natural state, these forests were probably burned frequently by ground-fires that maintained an open, park-like character. Now the trees form unbroken stands, through which any fires that do get started may spread more easily upward and thus kill not only seedlings but also mature trees.

In contrast to the changes in the forest since 1950, the clumps of bushes and prostrate trees at timberline on the slopes of Stanley Mountain to the right are nearly identical in both photos, suggesting environmental stability over the last thirty years. Moreover, the avalanche paths extending down through the forest on the mountain slope are similar in each view, although cloud shadows obscure them in Stewart's photo. The most prominent avalanche path, which extends from timberline to the valley bottom and crosses the road just beyond the point where the highway is masked by trees, provides the opening over which the view of the eastern approach was photographed.

The impression of few human activities in this landscape given by the previous picture is reinforced here. The tailing ponds of a molybdenum mine are visible in the leftmost canyon in the background, and the scar of an access road shows on the nose of the ridge, but otherwise the mountains appear little altered by development.

CONTINENTAL DIVIDE

Continental Divide

"Naturally, Father's road was nothing like the super-boulevard of today," remarked Lutie M. Bosley, daughter of the miner who built the first road over the 11,315-foot summit of Berthoud Pass in a mere sixty days in 1874. The occasion of her remark was a gala celebration in the summer of 1938 of the completion of U.S. 40, "hard-surfaced from coast to coast," as recorded in a history of the pass. The *Georgetown Courier* praised it as "the highest, most scenic and the shortest [transcontinental route] in the United States . . . the only motorway in the country that passes over the crest of the Rocky Mountains instead of around them." Governors from all the states through which the highway passed were invited, and on July 3 caravans of automobiles traveling from the East to the West Coast met at this point of the Continental Divide, "the top of the continent," and 3,500 people participated in activities which included a buffalo barbecue and a snowball battle between the East and West.[40]

Such a battle would have been possible on the June day when Stewart photographed this scene. On the snowless day of our July, 1980, photograph, a San Juan Tour Bus provided a meeting between a more distant East and West than those that were joined in 1938: the lodge set back to the left, out of range of the photo, hosted Japanese tourists feasting with chopsticks on a more recent specialty of the region than buffalo—stream-caught trout.

Originally traversed on foot by Indians, this route was eventually blazed by a trapper for his burro pack train and later improved for stagecoach travel. Although proposed by the engineer Edward Louis Berthoud as a route for the Union Pacific over the Rockies, the pass which bears his name was ignored by railroad officials, who preferred the less mountainous terrain of southern Wyoming. Early transportation was constrained by practical considerations, as reported in a history of Colorado: "The railroad engineers wanted easy grades, not stuff for poetry—not wildflowers, avalanches, alpine tarns, exotic fauna, and uplifting vistas."[41]

The builders of highways in the 1920s, however, according to the history of the pass, appreciated that the "presenting of this unique and model highway to the world" also "opened up a vista avenue of lodge pole pines, cavernous gorges, fascinating valleys, and panoramic views incorporating a number of our more important peaks whose hoary and lightning-scarred crowns are resting places for fleecy clouds."[42] In 1980, such attractions induced sightseers arriving by motorcycle, bus, and car to dash out and read the informational signs and snap a photo or two along the edge of this broad sweeping turn of U.S. 40, as they did also in Stewart's time. Through the lens of the camera, a photographer might be hard put to differentiate the vegetation of 1950 from that of 1980, for one can not only trace the same tree line across the canyon but also identify the same individual spruces "marching up the mountainside" at the left. Stewart's comment about the behavior of tourists here holds true today; crowds increase as the year advances from early to late summer, and our photo-taking was hazardous here where the passing lane for upgrade traffic ends, the slope eases, and the vehicles begin picking up speed.

The tourist season may actually peak with the onslaught of winter. Then these summer clouds, so typical of high elevations in the West, ever shifting and changing form and pattern, will give way to clouds ready to release as snow the moisture brought from the Gulf. Skiers and other winter sports enthusiasts will flock to the slopes of the "Snowy Range" from the most recent lodge built of native stone and logs in 1949, unless, of course, as in the first season of that structure, the skies unexpectedly withhold their bounty of white gold, leaving the hopeful skier with "everything a skier could want—except snow."[43]

The weather at Berthoud Pass frustrated us in 1980, but for quite a different reason. For three nights of stormy weather we camped in various campsites in the area and waited for enough sun to peek through the thick clouds to allow a good picture.

MEADOW AMONG MOUNTAINS

Meadow among Mountains

In a recent issue of a national news magazine, Colorado Governor Richard Lamm was quoted as saying, "There is no hyperbole that can do justice to how fast the West is changing. We are seeing a decade of change take place every month."[44] The spiraling growth to which Lamm referred is primarily the result of the explosive development of such energy sources as coal, petroleum, natural gas, and oil shale in the interior West. Here, about thirteen miles north of Kremmling, the landscape remains so unaltered over the last thirty years that Stewart's words describing the landscapes between Denver and Salt Lake City seem more appropriate: "It is all an empty land of far-scattered ranches and tiny towns. . . . There is still left one stretch of several hundred miles [of U.S. 40] not yet marred by billboards, not yet cluttered with more traffic than it was intended to carry, but offering a sufficiently good pavement and almost continuously varied and even magnificent scenery."

The ranch and surrounding slopes, little changed and certainly not "marred," still provide pleasing if not quite "magnificent" scenery. Along the stream, Stewart commented that "willows, alders, and other small trees grow in such profusion as completely to hide the water"; we found the same area supporting woody plants, although the growth seems thinner and few individual trees can be spotted that appear in both photos. The rancher may cut the trees for wood but permit them to grow back for subsequent harvests. The grass in the valley bottom away from the stream is still cut for hay, but our photo was taken too early in the year to show stacks of hay like those in the earlier picture. The "bushes encroaching upon [the meadow's] upper edge at the right" are still present, but the closest patch in Stewart's view has been cleared, revealing that the shrubs were growing on the slope and not in the meadow. The top of a wooden hay rack, a type called a "beaver slide," is visible just to the right of the closer truck in the 1950 photo, and it is identifiable in 1980 beyond the top of the fence post. Summer grazing by livestock has not altered the appearance of hillslopes above the meadow, most of which continue to be "covered with sagebrush and bunch grass." Colonies of aspen, particularly those on the dome-shaped mountain on the skyline, have the same borders in both views, suggesting a stability in the conditions which favor aspen in certain places and brush and grass elsewhere. The pattern of conifers on the sharp-peaked mountain is unchanged, although the shadow in the recent photo obscures the view. A small patch of dark conifers about halfway between the peak and valley bottom has neither expanded nor contracted; it likely is growing on an exposure of bedrock, as were the conifers on the hogback west of Denver. The wooden fence posts, for Stewart an indication that "wood is obtainable as a local product," have been replaced by a mixture of wooden and steel stakes, suggesting that ranchers today are as much influenced by desires for permanence or for style as by availability of local fencing material.

U.S. 40 itself, following the same alignment and having the same design, "swings in a wide semicircle, sacrificing distances for ease of construction, or merely following the old route." But the traffic on the highway seems to have increased somewhat since Stewart's trip, when "city-to-city trucks," "high-speed drivers," and "the more timid drivers" preferred the better-maintained surfaces and gentler grades of Highway 30 across southern Wyoming to the north. His photograph appropriately captures two ranchers' trucks and a single automobile. Today, even bulky, gas-hungry recreational vehicles find Highway 40 "sufficiently good pavement" for their purposes. Our photograph includes more than twice the number of vehicles in the earlier view. One auto pulls a small baggage trailer; a second struggles with a rather large house trailer. A skip-loader on the bed of a truck may be on its way to be used in the construction boom created by energy development farther west. The viewer is thus reminded that this "meadow among mountains" may in 1980 be an island in a sea of development, no longer the "typical" western landscape that Stewart saw it to be.

THE RABBIT EARS

The Rabbit Ears

"Sweeping in a fine curve," Stewart commented, "U.S. 40 rises on the final ascent to Rabbit Ears Pass, where it crosses the Continental Divide for the third and last time, at an altitude of 9680 feet." Maturing trees along the roadway of this stretch now bypassed by the main route caused us to stand farther back than did Stewart to gain an unobstructed view of the features in his photograph. A more evident difference between the appearance of the vistas results from ours being a snowless July scene, compared to Stewart's in nearly snowbound June.

The vegetation in this landscape illustrates the effects of both nature and people. The spruce forest on the slope in the background has thickened, filling in the low-growing nonforest vegetation, probably as a result of improved fire control. Fewer aspen seem present, also likely a consequence of a relatively long period free of burns. Even while humans have reduced the incidence of disturbance by fire, they have created new forms of disturbance, such as the cuts and fills of road construction that provide a habitat for sun-loving lodgepole pine. The highway cut on the far side of the pavement supported tree seedlings in Stewart's day, and these had matured into a dense stand of pole-sized trees by 1980. Their apparent vigor, moreover, suggests the ease with which trees become established in this environment on favorable sites, even though Stewart speculated that here "very severe wind action" resulted in tree growth that was "very scanty." The two lodgepole pines and the single spruce silhouetted against the sky at the right side of his photo are still recognizable and apparently healthy today. We do not see in the recent picture evidence of the "power of the wind action" which Stewart held responsible for the twisted branches of one of the two pines, but we can see the bent trunks of many of the young trees on the road cut where the wind undoubtedly deposits deep and heavy drifts of snow. The absence of trees or any but the scantiest of plant covers on the road shoulder in the foreground may be attributed to the grading of the surface by highway maintenance crews, the exposure to the sun's drying heat, or the compacted soil produced by vehicles pulling over to park.

The lack of trees on the shoulder is not surprising in Stewart's photo, which was taken when the paved road was, in fact, U.S. 40. Today, a reconstructed and relocated U.S. 40 crosses the ridge about one-quarter mile to the south, and the road in the 1980 photo is abandoned. The pavement appears to be in good condition, but the old highway's present function of providing access to a campground and small lake does not demand a paved surface. Undoubtedly it will eventually deteriorate.

The summit over which the new highway passes is identified as "Rabbit Ears Pass" even though it is not the same spot on the Continental Divide which is crossed by the old road. Like the summit of U.S. 40 on California's Sierra Nevada mountains, where a new alignment has replaced the old, the name of the "pass" seems to belong to the highway summit rather than to a topographic feature. Perhaps a low and well-defined gap in a high ridge would tend to anchor the name of a pass, but on the gentle terrain displayed by these photographs where a highway could be built almost anywhere, the name "Rabbit Ears Pass" may be as ephemeral as each generation of the road with which it is associated.

Stewart seemed impressed that U.S. 40 crossed the Continental Divide three times. Although most travelers are intrigued with this way of "dividing" the country, they seem to prefer a spectacularly mountainous crossing to one more subdued. Not surprisingly, then, rugged Berthoud Pass receives greater attention than the other two crossings by both picture-taking tourists and an English professor writing a book about the highway. Only a few people stop at the gentle landscape near Rabbit Ears Pass. And Muddy Pass, a few miles east of the Rabbit Ears, and barely recognizable as a drainage divide at all, is probably missed entirely by most travelers.

STREAM IN SNOW

Stream in Snow

When Stewart shot this eastward view, a few miles west of Rabbit Ears Pass, U.S. 40 was "a narrow and winding and comparatively primitive road" following the edge of the forest and passing to the left of the slowly meandering stream. Highway engineers have since widened, straightened, and further modernized the road to speed travelers to their destinations. Our view is slightly farther eastward than Stewart's because the fill for the relocated highway, visible in the left foreground, now nearly obliterates the scene from the original site.

Unlike the highway, the stream continues its lazy meandering course, seemingly in no great hurry to carry its load from the Continental Divide westward to the Yampa River, and then on to the Colorado. From there, in contrast, civil engineers have intervened to divert the life-giving water from its natural course and to transport it swiftly in linear pipes and canals to the thirsty lands of the Southwest. Besides its uses in the orange groves of California's Imperial Valley or the kitchen sinks of Los Angeles, though, this snow-melted water serves many purposes within the state of Colorado. Here in the Rabbit Ears Range, it may be sipped by deer, explored by trout, or admired by human visitors. Below the mountains, toward Craig, it may water the fields and pastures of farmers, who prosper when water is abundant but suffer when it is scarce, as described in the American Guide Series book on Colorado: "Water, like a magic fountain, has ever run through the day dreams—and at times, unfortunately, like an elusive phantom through the nightmares—of cowmen, sheepmen, beet growers, hay ranchers, truck gardeners, and general farmers."[45]

Except for seasonal differences, the stream and most of its setting seem little changed or disturbed by time or humans. Stewart believed that in a few years most of the slope on the right would be "covered with a clump of lodge-pole pines," yet the forest of lodgepole and spruce has only slightly thickened and invaded the meadow. The advance of trees noted by Stewart was perhaps a result of heavy sheep grazing before 1945. This grazing may have opened up the thick sod to allow tree seedlings to germinate for some years until the grass and shrubs returned to their former condition. In our photo, the stream corridor is covered thickly with herbaceous and shrubby willow growth, and it may be this unbroken cover that further halts the invasion by lodgepole.

The changes we observed taking place in the vegetation in 1980 were rather subtle, and were the result of a different type of human disturbance, the highway itself. A close look at the photos reveals that lodgepoles are thickly invading the old road and road cut, to the left of the new roadway. Behind us, recovery of vegetation on the old road cut was visible, but slower, with some herbaceous growth and small trees. The new fill below us was being covered by grass, and to our left the new guardrails sheltered several one-foot tree seedlings. We wonder if a future photographer will find the view blocked by a wall of trees, or perhaps a mountain of new fill. Will the traffic become even heavier, drowning out the twittering of the pine siskins and white-crowned sparrows and the buzz of the hummingbird? Or do the abandoned telephone poles crossing the stream, with wires drooping lifelessly to the ground, foreshadow that the highway here, as at old Rabbit Ears Pass, will by that day be only a side road?

OATFIELD

Oatfield

"Drop two thousand feet," Stewart began, "shift from June to September, and the deep snows and louring clouds of the high country are transformed into the hot sunlight of a recently-harvested field of oats." Wait thirty years, we can add, and the oatfield is changed into an expanse of irrigated hay.

The view is a few miles east of Craig, in the relatively dry environment of western Colorado, where mountain ranges to the east block moisture from the Gulf of Mexico, and mountains to the west reduce moisture from the Pacific. The climate "approaches a desert condition" as reflected in the "scanty vegetation on these hills," Stewart commented, and in 1980 the plant cover appeared little different, except maybe somewhat denser. The large tree beside the highway in 1950, however, was gone in 1980.

The environment is too dry for most agriculture. Stewart suggested that the "fine crop of oats . . . is to be credited partly to an unusually rainy year and partly to the moist character of the low-lying land near the [Yampa] river." The 1980 hay was dependent upon neither the precipitation which fell directly on the field nor the river maintaining the natural soil moisture. Rather, the field was irrigated by two ditches that probably took water from the Yampa upstream and which ran more or less parallel to the highway; they were hidden from view by the tall grass. Water seeping through or spilling over the sides of the ditches flowed slowly down the slope from left to right. The field at the time we took the photograph was soggy underfoot, and our shoes became soaked on the hot afternoon we waded through the grass to locate the photo site. We could easily believe that the abundant water and heat allowed for multiple crops of hay even though the field was at the relatively high elevation of 6,000 feet.

In a sense Stewart was fortunate to have encountered a field of oats where he did. The 1950 Census of Agriculture reports only 1,800 acres of oats in the three million acres of Moffat County, through which U.S. 40 passes for one hundred miles from east of Craig to the Utah state line. The odds of our finding oats were even less: only 200 acres were planted in Moffat County in 1974. It is not surprising that we did find hay here because that crop is the primary irrigated crop in the county today, rivaled only by wheat, which is planted on unirrigated uplands such as those farther west. Moffat County is not, however, a cornucopia of agricultural products; its 60,000 acres of harvested acreage, about the same as in Stewart's day, represents only 1 percent of the cropland in the state of Colorado. Low rainfall continues to limit agriculture as it did in Stewart's time.

The highway has been straightened, widened, and perhaps raised somewhat since Stewart drove through. A few curves have been eliminated outside the photo view (to the left), but the visible stretch of roadway seems to be built on the old alignment. The road cut in the background is larger, and the poles of the telephone line have been relocated to cross the hill at a higher elevation. Traffic on this segment of U.S. 40 is busy with trucks of commerce. Stewart's photo recorded a single truck for transporting livestock, while ours captures a paneled delivery van, a flatbed moving an oil-storage tank, and a truck hauling aggregate. Even in this pastoral setting, reminders of the development of western energy resources are never far away.

SHEEPHERDER

Sheepherder

We had some initial difficulty finding this site, "a few miles west of Craig," but succeeded by identifying "the rolling hills in the background," with their unchanged pattern of brushy vegetation. In the more recent photo, the notch in the slope of the rise near the center of the view indicates the right-of-way of the old highway, and its former bed can be traced toward the foreground by following the band of light-colored bur clover which has invaded the disturbed ground. U.S. 40, today relocated slightly to the south, is no longer "at its lowest ebb—narrow, sharply curving, lacking good shoulders, unfenced, rising at a steep grade." The present highway, enclosed within a steel fence, has twelve-foot lanes and seven-foot shoulders, the latter much reduced to allow for a separate passing lane as the pavement rises in a long gentle straightaway toward the low crest beyond the hill.

Stewart considered not only the highway but also the telephone poles and house to be "far from . . . truly up-to-date." Today's poles carry fourteen lines, rather than ten, and a cable is buried along the same alignment. Additional utility lines have been installed across the road. The house is barely visible through the maturing elms, which, together with a roof of asphalt tiles instead of corrugated iron, give it a less stark appearance than in the earlier view. Still, it is hardly a "blue-ribbon farm"; an old car and farm machinery rest foresaken among added outbuildings and a stack of hay.

This "mingling of old and new" was a theme Stewart perceived here. His discussion centered around the changing role and mode of sheepherding, and we noted that the slow evolution of agricultural use has continued. Stewart echoed the sheepherder's opinion in questioning the wisdom of the owners of the house "growing spuds" on this "marginal land." We saw no potato field here or nearby. Actually, only ten acres of potatoes were grown in this county in 1950, and a similarly trivial acreage is planted today. However, although the sheepherder as well as Stewart thought sheepherding to be a wiser use of this dry land, it has been the sheep industry that has declined sharply, rather than other agriculture. The number of sheep in Moffat County has decreased one-third since the earlier photo, from 151,000 to 110,000 head. Cattle, on the other hand, have increased twofold. Cattle-related operations seem to have edged out both potatoes and sheep here, since hay is now grown adjacent to the house, and a new large barn has been raised on the old right-of-way behind the camera site.

Cattle, indeed, are the main domestic animals one sees while traveling along U.S. 40 today. Few sheep dot the hillsides, and the modern farmstead is not surrounded by "Old MacDonald's" barnyard fowl. Although Stewart photographed one horse here and others elsewhere, he astutely wrote concerning the mechanized equipment being used at "Hayfield" that "one can drive the whole length of U.S. 40 through much of the country's richest agricultural land and see only an occasional horse at work in the fields." We saw none at work in the fields, and surprisingly few being used for pleasure riding.

An animal that has not been replaced through either changing economics or modernized machinery is the dog, in Stewart's time used by the sheepherders to protect their herds from coyotes and mountain lions. Coyotes are too small, and mountain lions are now too rare, to be a threat to cattle, but a "sheepdog" still guards the mobile home above the old road right-of-way; it even made "a few threats to the photographer's legs," just as those in the earlier photo did in their time.

The sheepherder's "home-on-wheels" greatly interested Stewart as the then-current product of the evolution of the early Conestoga wagons that helped to blaze the first trails through the West. The traditional canvas cover of the wagon, he pointed out, contrasted with its modern "running gear . . . taken from an automobile truck." We see the "mobile" home as a more recent descendant of the American covered wagon, here more home than wagon, with "running gear" probably replaced entirely by a concrete slab foundation.

The controversy over the use of this land when Stewart stopped by was between farmer and sheepherder. Today, eastern Colorado, like much of the interior West, is faced with continuing controversy over development of energy resources. Just a few miles toward Craig from this location, for example, coal strip miners threaten to change the natural setting much more rapidly and dramatically than highway, sheepherders, farmers, and cattle grazers combined.

ROCK WALL

Rock Wall

The description of western Colorado included in the American Guide Series book seems as appropriate today as for either 1941, when the book was published, or for 1950, when Stewart stopped for his photograph. "West of the chain of parks the mountains shelve off into the mesa country that stretches away toward the purple sage flats along the Utah boundary, a beautiful but often sterile land inhospitable to man and beast."[46] If "inhospitable" can be associated with "emptiness," our 1980 view presents an even less kind landscape than Stewart's. He showed at least three autos on the highway, and one person at a mailbox. The recent photo reveals no travelers, the only automobile being our own. It was easy to park on the pavement and on the wrong side of the road, because this stretch of highway is abandoned. No people or "beasts" are visible anywhere.

The "isolated ranch house located on the sideroad" to the right of the photo and in which the man at the mailbox undoubtedly lived is now gone. The number of farms in the interior West—as in the Appalachian hill country—has declined since Stewart's day. Moffat County, Colorado, had 340 farms and ranches in 1950, but only 249 in 1974. The number of harvested acres is about the same, and small sheep operations have been replaced by larger cattle herds. These facts suggest that the average farm size has increased greatly, a widespread trend in America. It is a trend that often results in an emptying of rural landscapes.

The reconstruction of the highway allows people to speed across this broad flatland, probably accentuating the feeling of its inhospitableness. The highway has been straightened and relocated far to the south in order to avoid some dissected terrain to the left of the photo view, although it still crosses the ridge in the distance at the same point the old highway did. The pavement of the original highway seems to end abruptly in the distance, but actually a side road extends to the left at a right angle at that point to join the new U.S. 40. The fence paralleling the old highway in the foreground, not present in Stewart's photo, was probably erected before both the road and the farm were abandoned.

The vegetation continues to support range livestock, even though, as Stewart noted, the climate is sufficiently dry that "desert conditions are approached." This great expanse of open land is covered by annual grasses, greasewood, and some sagebrush, as it probably was in Stewart's day. Even a small area of browned grass in the right foreground is visible in both photos. Stewart speculated that the woody shrubs may have increased, and grasses decreased, prior to his visit: "Fifty years ago the flat in the middle distance might have been all grass-grown instead of being largely covered with a growth of bushes that are useless for feed." However, judging from descriptions of the vegetation in early journals, woody plants were an important part of the plant cover in the interior West before the arrival of domestic cattle and sheep.

The two junipers on the slope in the foreground have grown in both height and breadth, although the amount of growth may seem meager to anyone accustomed to tree growth in humid environments. The line of planted trees just in front of the massive rock wall has suffered, with only a few trees surviving. Those that are still alive include the two largest in Stewart's photo, probably because they had the best-established root systems and thus were able to maintain themselves after the ranch was abandoned.

We wonder how this scene will appear after another thirty years have passed. Range livestock will likely continue, although changes in economic forces that would prompt resettlement of the ranch are hard to imagine. Lack of water will probably still limit intensive agriculture. The area is too remote to allow the building of homes, and the landscape seems to offer little to the typical recreationist. The development of oil shale deposits in similar country just south of here would encourage human activities throughout western Colorado, but how much or what sort of change might occur here cannot be foreseen. Perhaps U.S. 40 will still be crossing an "inhospitable" but "beautiful" landscape "that stretches away toward the purple sage flats along the Utah boundary."

STATE BORDER

State Border

We photographed this scene of U.S. 40 at the state line between Colorado and Utah in three successive years. In 1978, no signs announcing the state borders were visible; by 1979 the state of Utah had erected the large modern sign seen in our photo, but in 1980 that sign was gone. We found the posts broken off close to the ground, and pieces of the wood scattered about in the brush. An errant driver probably had eliminated the evidence of the official end point of one state and beginning of another.

Nonetheless, on each of our visits many signs extended away from the state line in both directions, as at the boundary separating Maryland and Pennsylvania. Here, within the first two-tenths of a mile into Colorado, five signs line the highway, identifying the county, welcoming visitors to the state, and announcing the highway number, distances to upcoming towns, and the speed limit. The traveler entering Utah drives by an almost identical array of signs, only one of which is visible beyond the "Utah Welcomes You" pronouncement. The careful observer will also note that the width of the pavement changes slightly at the state line, with Colorado having a wider paved shoulder than Utah. The placement of the white line marking the edge of the shoulder is accordingly adjusted. Overall, though, the differences in highway regulations and standards are trivial, and Stewart's comment about the meaning of state borders seems just as true today as in 1950: "In the highly federalized United States, since the Civil War at least, state boundaries have . . . more a sentimental than a practical significance."

The author Joel Garreau has recently given formal support to the idea that state lines have little meaning by identifying eight regions of the United States, each unified by both natural and cultural characteristics and each with "its own way of looking at things." State boundaries rarely coincide with the borders of his regions. The state line between Colorado and Utah, for example, is well within the "Empty Quarter," which includes most of the Rocky Mountains and northern intermountain West. Garreau argues that a distinguishing characteristic of this region is the ownership of most of it by the federal government; that happens to be true of all of the land in the photo view. Regarding the exploitation of resources in the Empty Quarter, he suggests not entirely with sarcasm that the region "is being chewed up and spit out in order to light our lamps and power our air conditioners."[47] Whether the exploitation of this region for its resources is greater than that of the "Breadbasket" for its food or of the "Foundry" for its heavy industry, however, may be open to question.

Stewart stressed the absence of major differences across the Colorado-Utah border and thereby lent support to the notion that western Colorado and most of Utah are more alike than different. The speed limit in 1950 was sixty miles per hour in both states, the highest of any stretch of U.S. 40 at that time. In 1980, in a more "federalized" United States, the legal maximum speed had been nationally standardized at fifty-five, much to the chagrin of many western states. In 1950, the gasoline tax was 7½¢ in Colorado, higher than any other state along U.S. 40, and 5½¢ in Utah. When we passed through, the price of gasoline was regulated by the federal government, although the taxes had risen to 11¢ in Colorado and 13¢ in Utah. We paid $1.18 a gallon for gasoline in 1980; Stewart paid about 27¢. "As far as the country itself is concerned," Stewart commented, "—barren and wholly without habitation—the shift from one state to the other may be called a purely theoretical transition."

Even if the borders between states have little practical significance, they are often places where states try to impress the traveler. Here, Colorado has done much more to present a positive "first impression" than Utah. A plaque dedicating the highway as a "Blue Star Memorial Highway," commemorating Americans who died in World War II, was emplaced in a monument of red rocks just to the left and behind the camera position. A thoroughly modern touch was the message "Tomas and Lupe" sprayed neatly with yellow-green paint across the plaque. Next to the monument a low barbecue, also constructed of reddish rock, rested unused with some paper jammed beneath the grate and a few weedy plants growing in the fire pit. A ponderosa pine had been planted and nurtured in this unlikely spot, probably to give credence to the image of Colorado as a mountain state. In 1978, the tree was about twenty feet tall and apparently healthy; in 1979 it was dead but standing; and in 1980 it had been cut down. Two other planted trees, a small elm and a smaller juniper, were seemingly growing well.

ROOSEVELT, UTAH

Roosevelt, Utah

The Uinta Basin, in which the small community of Roosevelt is situated, was described in 1861 by a Mormon exploration party as "measureably valueless . . . except to hold the world together."[48] In the same year nearly the entire basin was designated as the Uintah and Ouray Indian Reservation by President Lincoln. A decade later, livestock ranchers settled in the valley, to be followed by farmers. Roosevelt was established in 1908 and became a trading center for these local agriculturalists. By 1950, Stewart believed that the community, "being the best place between Vernal, 27 miles to the east, and Heber, 99 miles to the west, at which to find a decent lodging for the night, or eat a meal, or buy a new tire, or have a little garage work done, or amuse yourself by an evening at the movies," had come to depend primarily upon the highway for its income. Two hotels, two cafes, and a grill welcomed the traveler. Stewart found the false front on the Roosevelt Hotel and the wide streets typical of the traditional western town. He also noted that although Roosevelt was a Mormon town, the only evidence of Mormon culture was the deep gutter at the right of the photo. He described U.S. 40 as "here only a narrow ribbon through primitive country," with "dirt or gravel roads and streets [leading] into it from both directions." Overall, he described the town as "somewhat dilapidated and run down," with "the only sign of extravagance" being the "showy and new electric street-lights."

In 1980, in general, our impression of Roosevelt was one of prosperity. It is still a trading center for the ranches and farms, as well as the nearby Indian reservation. Its income from travelers along the highway continues to grow in importance as rising gasoline prices encourage tourists from Colorado and Utah to remain close to home. A nearby promotional sign reads, "You do not have to drive far to have a vacation in Dinosaur Land." Roosevelt is probably still a popular resting place, since the 1981 American Automobile Association Guide Book lists only three motels between Vernal and Heber, one in Duchesne and two in the Roosevelt area, including Bottle Hollow, a resort which is seven miles to the east and operated by the Ute Indians. Many businesses have taken on new false fronts, some of them more European than western, such as the painted wooden slats crisscrossing the stucco of the Roosevelt Hotel and the domed and scalloped awning on the second story of the dress shop on the left. The Western Wear store, appropriately, has adorned itself most extravagantly with a rustic wood-shake overhang. A more cosmopolitan air is added by the changing of the Roosevelt Cafe to Dana's Casa Mexican Food. Several other "signs of extravagance" are the again-new street lamps, concrete planters of petunias and marigolds beside painted litter barrels, overhead traffic signals, and electric and telephone lines out of sight in alleys behind the buildings. A new bank has been built to the left of the picture. The town, in fact, was bustling with activity when we walked along these blocks. The cafe in which we stopped for coffee barely had room at the counter for two weary travelers.

The busiest activity, though, was not along these streets, but on them. A new industry—natural gas and oil operations—is bolstering the economy of this land once viewed as "measureably valueless." The 1980 local telephone book contained twenty-nine yellow pages devoted fully to listings of oil field services and well drilling; Plateau, Inc., a petroleum refinery, has an address on the west side of town. The "boom spirit" which Stewart believed prompted the sixty-nine-foot width of the street may finally be realized. The four traffic lanes, left-turn lanes, and parallel parking which replace the former two lanes and diagonal parking did not thin out the traffic enough for a casual photographer to run to the center line for a quick shot without the aid of signals from the light behind him and his wife beside him. It took careful timing and coordination for us to dart out between both onrushing commercial trucks, which seemed to be in a hurry to "hold the world together" in this period of perceived energy shortages, and passenger cars, seemingly rushing to see that world before the gasoline pumps run dry.

The long waits on the sidewalk for a break in the traffic allowed us time to study this end of the block more closely. Visible Mormon influence continues to decline. A vacated store with Mormon religious articles in the window in 1979 was completely empty and for rent in 1980. The new front on the Roosevelt Hotel seemed truly false. Up close, it appeared largely vacant and run-down, with windows framing torn screens and ragged drapes. A sign advertised, "Weekly and monthly rates only." For only a few minutes, the "boom town" reminded us of the fate of so many of the Western towns dependent upon mining in earlier days.

PLATEAU COUNTRY

"West of Duchesne, Utah," Stewart commented, "U.S. 40 passes through characteristic scenery of the Colorado Plateau. . . . The valley is moist, being along the course of Strawberry Creek. Cottonwoods and sycamores grow along the stream, and there is even—with the aid of irrigation—some farming." In 1980, the bottomlands in Stewart's picture were beneath the waters of Starvation Reservoir. Our photo is west of Stewart's, but the view is in the same direction—east—and the cliffs in the distance are identical in both pictures.

The reservoir results from the damming of the Strawberry River, apparently promoted from "creek," which flows east from the Wasatch Mountains across the Uinta Basin, where it collects runoff from the south slope of the Uinta Mountains before joining the Green River southeast of Roosevelt. Completed in 1970, the dam was built by the federal Bureau of Reclamation as part of the Central Utah Project, which allows more extensive irrigation of agricultural land in the Uinta Basin than was possible in Stewart's day. This reservoir and two others in the project area together provide water for about 12,000 acres, planted mostly in forage crops for livestock and in grains—just a little over 1 percent of the one million acres of harvested cropland in Utah, but valuable to the farmers involved. The lack of great volumes of water is suggested by the surprising shallowness of the lake: rock formations that break the surface of the lake in 1980 can be seen to be close to the valley bottom in the 1950 photograph.

A secondary explicit function of the reservoir is recreation. Several miles of the U.S. 40 of Stewart's time serve as access to the lake. Rough and poorly maintained, this segment extends from a relocated and modern highway which crosses dry uplands and passes through a meager juniper forest completely unlike the moist environment in Stewart's photo. Barriers have been erected to warn drivers unceremoniously of the impending submergence of the pavement. U.S. 40 beyond becomes a boat ramp.

In the 1980 photo, a group of campers have settled their trailer near the water's edge, where they have easy access to the lake for water-skiing. The shoreline, whose rather barren slopes would discourage us from camping there, is in marked contrast to the expanse of cottonwoods and sycamores that formerly graced the valley bottom. This lack of tree growth around the edge of the lake is typical of reservoirs in arid regions, where the fluctuating shoreline inhibits the survival of woody plants. Here, in addition, the water level abuts upon hard rock, and thus few sites exist which permit the rooting of trees or large shrubs. The campers in the photo have responded to the absence of natural shade by erecting a large striped canvas as protection from the blazing sun. Completely unsheltered from the heat are two wooden outhouses set back on the flat to the right. These twin structures are, in a way, replacements for the two "genuine" early log cabins of "bare-subsistence level" farms which Stewart noted in his photo.

A large oil-storage tank, visible on the horizon atop the closer mesa, also sits in the sun, presumably unaffected by the heat. Its presence is a reminder of the importance of the Uinta Basin as a producer of petroleum and natural gas. Old U.S. 40, in addition to allowing boaters to reach the reservoir, also provides access to production fields for these fossil fuels.

Conservation groups often oppose drilling for petroleum in wild landscapes, but they usually are most adamant in their opposition to dams and the reservoirs dams create. Perhaps the parallel between human life and a living, freely flowing river—arising from small, energetic mountain streams, developing into more powerful but less boisterous main channels, and finally slipping quietly beneath the ocean depths—is strong in the human mind. Beyond the Green River on the Colorado, proposals for dams in the canyons of southern Utah and northern Arizona were major national conservation issues of the last three decades. Although we know of no strong opposition to the construction of Starvation Reservoir, it stands as a good example of the mixed blessings that come from such projects: while it has clearly aided farmers and water-skiers, we think that it has not ennobled or beautified the landscape.

BEAVER DAMS

Beaver Dams

On a small stream beside U.S. 40 where it climbs out of the Uinta Basin and enters the westernmost ranges of the Rocky Mountains, Stewart was particularly intrigued with, and devoted much attention to, the twin beaver dams. When the beavers first arrived at this spot, he speculated, "the aspen forest must have grown clear down to the stream. . . . Each night, going to the nearest point at which aspen was available, the beavers cut their trees and obtained their food. Gradually, however, in the course of years, they were forced to cut farther and farther from the edge of the water. . . . At the present time, the beavers have possibly reached a critical point where food-supply is so far from the water that they can reach it only at considerable risk. . . . A civilization is about to fall."

Stewart's prediction was astute. Highway reconstruction eliminated the slope on which Stewart stood and forced us to take our photograph lower down and perhaps closer to the dams, where the growth of willows blocks our view of the small area of open water remaining. We searched without success for signs of active beavers. The aspen so heavily cut-over in 1950 have sprouted vigorously, and a young forest is reestablishing itself, implying that the timber-harvesting beavers are gone.

Stewart was also correct in suggesting that "small meadows are gradually forming behind the dams." By 1980, most of the pond areas had filled with silt and had been invaded by moisture-demanding plants, including grasses and sedges. Willows had taken root both along the dams, as can be seen by the line of shrubs extending across the valley in the recent photo, and within the pond areas where sufficient soil had accumulated. In addition, a young spruce tree had become established just beyond the lower dam. Yet, enough open water remained to allow a female duck, probably a mallard, to nest and hatch her eggs; we saw the family swimming nervously on the lower small pond.

Stewart commented about the importance of the dams for "water control": "The impounded water leaks out slowly during the dry season, thus maintaining the flow of the streams, and the half-empty dams prevent the downward rush of the sudden flood waters." Like Starvation Reservoir, built by even busier but no more clever "beavers," these small dams helped to "even-out" the flow of the stream. Does the water flow below the beaver dams fluctuate more now that the ponds are mostly filled-in, or do the wet meadows continue to "leak out slowly" and thus help maintain a more constant streamflow?

The beaver dams may continue "to provide the chief interest," but we found other features here, mostly related to animals, also worthy of note. The fence is no longer "of local juniper wood, the small posts of which require to be set thickly," but rather of the white-topped steel stakes which line many newer highways in America. In 1980 at least, the lower wire of the fence was barbless mesh designed to constrain sheep, such as the large flocks we observed west of the photo site. We were not sure if the fence was intended to keep sheep off the highway or off the land containing the beaver dams; the white sign on the power pole announces that the land behind the sign is owned by a private hunting and fishing club which, we thought, may not want sheep grazing in its wildlife habitat.

The environment created by the beavers was fine for other wild animals, even if the beavers themselves seem no longer present. Yellow warblers and song sparrows sang cheerfully from the willows, while tree swallows and a flycatcher caught insects in the air over the ponds. We saw a weasel darting away from us through the brush beside the fence. And, just as we were leaving, we caught a glimpse of a small mammal swimming in the water.—A muskrat? Or have not all of the beavers realized that their "civilization" has "fallen?"

VALLEY IN THE WASATCH

Like Roosevelt, Heber City was a Mormon settlement. But this landscape, three miles north of Heber City and within the Heber Valley, much more strongly reflects the affinity of the Mormons for water, trees, and stability than do the streets of Roosevelt.

Along the horizon to the southwest across Heber Valley rise the peaks of the Wasatch Range. At their base, from right to left, runs the Provo River, identifiable by the thick growth of cottonwoods on its banks. The Utahan attitude towards water, according to the American Guide Series, has been that "it is unnatural that rivers should waste into the sea, just as it is unnatural that farmers should mature crops by rain alone. Rivers should be dammed at canyon mouths and their waters carried in canals to the thirsting land."[49] The drainage here is part of the Great Basin, and would never "waste into the sea," but Mormons have still improved upon nature and dammed the river at the gap to the left of the photo. The light-colored area is Deer Creek Reservoir, which stores much-needed water for cities and farms below the mountains to the west. Utahans show the same practical conservative attitude towards the land in general. Until recently Utah was the only western state without United States Forest Service wilderness. All of the Wasatch visible in this view is part of the Uinta National Forest and includes some of the highest elevations and most rugged topography of that range, and yet none of it has been designated as wilderness or even as a wilderness study area. One concession was to establish the Mt. Timpanogos Scenic Area in 1961, representing 10,570 acres, visible in the upper right center of the photo.

"Rich meadows" stretch through the valley today as they did in 1950, irrigated by seepage and flood irrigation from canals. Only one of the two barley fields remains, however, with the rest of the agricultural plots in hayfields or pasture. A herd of cattle can be seen, though with some difficulty, through the limbs of the tree partially hiding the former barley field. A few cottonwood or willow trees still mark the fence lines; some of those in Stewart's time may have died naturally, as have at least two in the recent photo, whereas others may have been grubbed out as they were in "Hayfield," Kansas, to allow more efficient, mechanized farming. Modern farming techniques, requiring less human labor, have probably been the reason for the decline in the number of farms in Wasatch County by about one-third since Stewart's time. Otherwise, agricultural land use in the county has been stable, with irrigated acreage increasing slightly as harvested acreage decreased a similarly small amount. Numbers of cattle, sheep, and acres of hay have fluctuated little. About 600 acres of small grains like barley were lost, as evidenced here, but an equal number were planted in wheat.

Stretching across the foreground is the strongest reflection of Mormon culture in any of our photographs from Utah. In our view, the thicket of Gambel oak, fed by seepage from two parallel irrigation ditches between which we stood, has grown so tall that it hides much of the farmstead below. Stewart wrote that the farm "displays much of Mormon solidity," and he particularly noted the sheltered brick house, the well-built and painted barns, and the brightly shining corrugated metal roof of the shed. All remain standing and well maintained today. The house, barely visible through the base of the dead branches of a cottonwood, has been painted white. Here the "sun reflects brilliantly" from an additional barn below the power pole. On this June day, it was filled with bales of hay. These were probably part of the first harvest from the fields across the highway.

The 1980 highway is relocated and slightly elevated from its 1950 alignment. The former road right-of-way can be seen at the end of the pitch of the roof of the farthest barn. Straightening and widening were apparently not the reasons for this new road-building. A close look at Stewart's photo shows much standing water at each end of the farm, perhaps from heavy rains, which might have presented a drainage problem for the adjacent highway. Unlike the Mormons, highway engineers moved not the bed of the water, but the bed of the road.

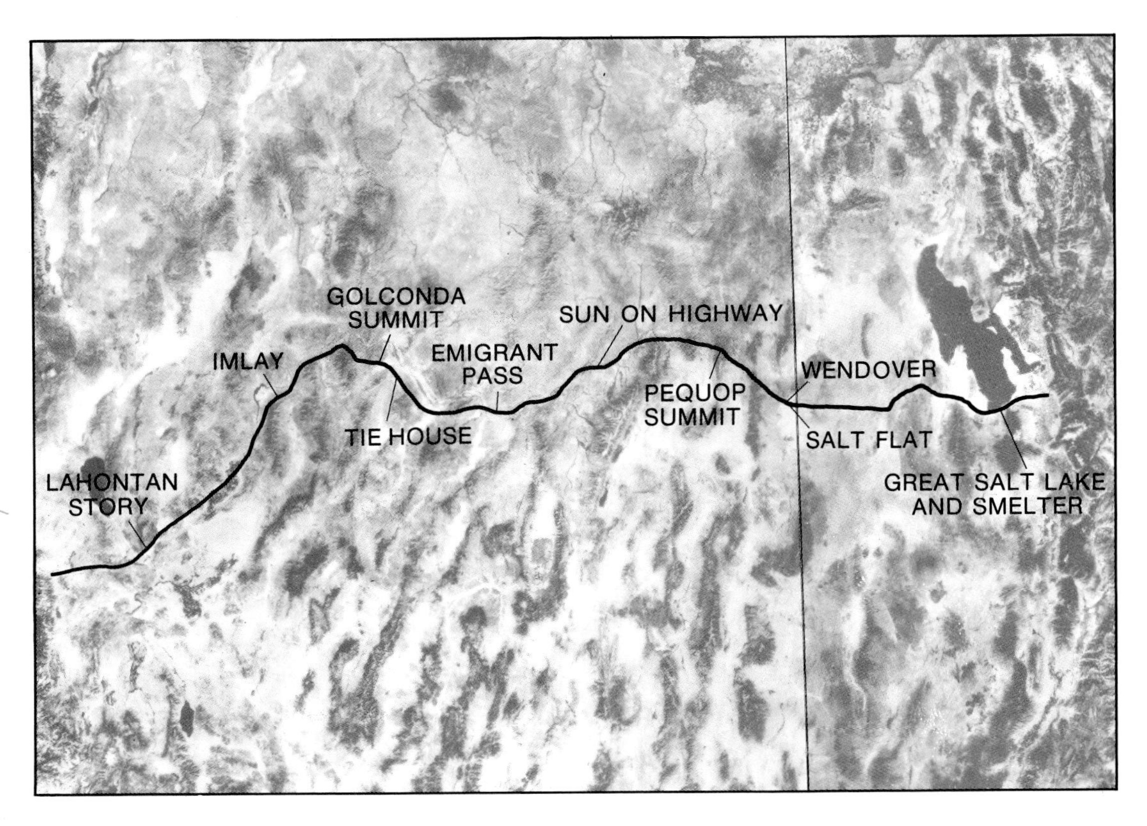

SALT LAKE CITY TO RENO: THE DRY SECTOR

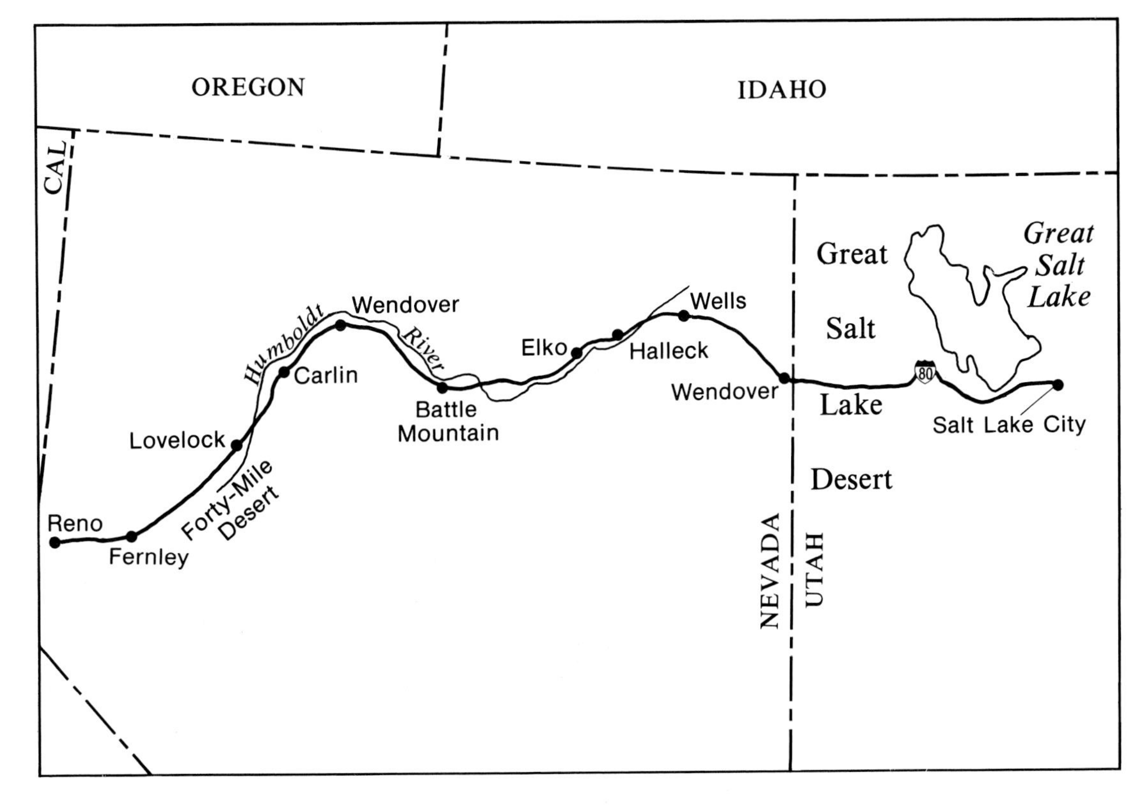

Salt Lake City to Reno: The Dry Sector

THROUGH WESTERN UTAH and all of Nevada, U.S. 40, now designated Interstate 80, crosses the Great Basin, a vast region whose streams never reach the ocean. It is dry country, full of sun and solitude, salt flats and sagebrush slopes. West of Wells, Nevada, the highway closely follows the California Trail, used by overland emigrants bound for the Golden State. The route has thus been the major east–west way of travel through the deserts of Nevada for nearly a century and a half, largely because it follows the Humboldt River, a stream that disappears into the Humboldt Sink north of Fernley.

Like early fatigued travelers, who had already been months on the trail by the time they reached the Great Basin, contemporary tourists likely find the desert crossing monotonous. After the bizarre salt flats of the Salt Lake Desert, the road traverses a seemingly endless expanse of brush and rock, occasionally broken by fields of irrigated alfalfa or small but busy towns. Yet, for Stewart this segment through Nevada had a special appeal. He was intrigued with its historical importance as the modern equivalent of an emigrant road. Besides two books on the California Trail, Stewart wrote still another book, *Sheep Rock,* set in a locale west of Winnemucca. In *U.S. 40,* he seemed particularly impressed with the broad sweep of valleys ending against distant mountains, for most of the photos he used have this characteristic. In contrast, he seemed uninterested in more intimate views of Nevada, such as small towns or crossings of the Humboldt River. Moreover, in his introductions to other segments of the highway, Stewart frequently used adjectives like "pleasant," "beautiful," and "magnificent" in referring to the environments through which the road passed, but in none of those essays did he describe features with such apparent affection as he did for Nevada: "Scenically the region is scarcely second to any. Jagged ranges of mountains, sometimes snowcapped, trend north and south, and between them lie desert valleys, majestic in empty distances. Mirages distort the sky-line, and dust-devils blow on the salt-flats. Though the land itself is often a gray monotone, the light and color of the desert fill the atmosphere, and the clouds offer infinite variety."

The same comments about the Nevada landscape could be made today, the land in general having changed little over the last thirty years. The road, of course, is no longer "a broad two-lane highway," as Stewart described it, but a full freeway. The final segments of the interstate, bypasses of the major towns, are being completed as this is written. Stewart seemed to anticipate this transformation of the highway, for even in 1950 he commented that U.S. 40 across Nevada had "nothing of that lonely quality that it displays in western Kansas, across most of Colorado, and in eastern Utah. . . . Already its two lanes are beginning to seem crowded. . . . Through all the hours of the night the buses and the long-distance high-speed trucks roar through."

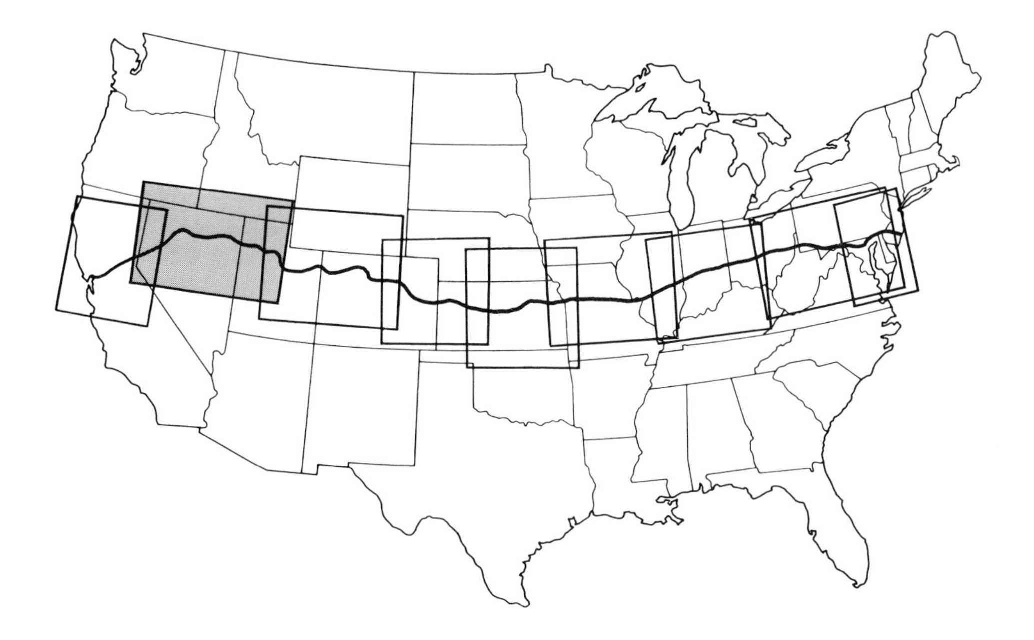

GREAT SALT LAKE AND SMELTER

Great Salt Lake and Smelter

When Stewart stood on top of Black Rock, seventeen miles west of downtown Salt Lake City and at the edge of the Great Salt Lake, he said that the former island had been connected to the shore as a result of receding waters. In 1980, the water level seemed higher, judging from the distance between the shoreline and the railroad. But Black Rock was not again an island, because earth and rock, such as that in the foreground of the recent photo, had been dumped here to make the connection more permanent.

In Stewart's time, the linking of Black Rock and the mainland may have been strengthened with artificial fill in order to enhance the recreational use of the former island and its bordering beach. The American Guide Series book on Utah reports that, in 1941 at least, tourists could gain admission to "two privately-owned lake resorts" for an "admission of ten cents"; "bathing" and "dancing" were available.[50] Stewart mentioned only bathers on the beach. When we stood on Black Rock we were impressed with how few recreationists were using the site. One shirtless and shorts-clad man, his skin tanned darkly, walked slowly along the water's edge, briefly raising small swarms of brine flies as he progressed. A panel truck and trailer, seen toward the right of our photo, was parked beside the beach, immobilized by the soft sand. While we watched, a rescuing tow truck, also visible in the picture, drove out the unpaved road paralleling the beach until it too became temporarily stuck. The truck driver eventually freed his vehicle, and presumably was then able to help the careless tourist.

If recreational uses of the area in the photo view seem lessened since Stewart's day, the tendency of transportation facilities to be wedged between the Oquirrh Mountains on the south and the Great Salt Lake on the north is as true as in 1950: "All communications are squeezed into the bottleneck between the steep slopes and the water." But some of the facilities reveal change. The railroad is still present and busy with long-distance hauling, but the "piggy-back" truck trailers on the railroad flatcars are a decidedly modern touch. The pipeline which Stewart noted has been enlarged considerably and relocated, requiring the grading back of the cliff at the extreme right of the photo. U.S. 40, which merges with Interstate 80 east of Salt Lake City, crosses the railroad on new bridges and parallels the railroad and pipeline to the right. The road that was U.S. 50 in Stewart's day came from the south to join U.S. 40 just west of here, but today is not visible because of the freeway fill; today's U.S. 50 follows a route 100 miles to the south.

Behind all the transportation facilities, the large copper-smelter rests against the base of the mountains. It continues to process the ores mined farther south in the Oquirrhs, notably at Bingham Canyon, the world's largest open-pit copper mine. The huge new stack was undoubtedly built in order to allow better dispersal of smoke, which is so prominent in Stewart's photo. Only two of the three stacks which Stewart described as "high" remain, and the low buildings at the heart of the smelter are much changed, perhaps reflecting the altered technology required to extract copper from ever-leaner ores.

Stewart commented about the effects of the gases released by the stacks: "Heavily charged with sulphur dioxide, this smoke has completely blighted the growth on the nearby mountains." About ten years after Stewart passed by, a local resident, Paul Rokich, began a personal effort to reestablish the natural vegetation on these slopes by trespassing on the land owned by Kennecott Copper Corporation to plant trees. According to an article in *National Wildlife,* "Parking his car on the edge of Highway 40, in the dead of the night, he toted seedlings up the steep hills for 16 hours at a stretch . . . risking flash floods, snake bites and rock slides, and gradually the fruits of his labor began to flower." Subsequently, hired by Kennecott Copper to continue his planting activities, Rokich claims to have "personally planted some 10,000 trees, 5,000 shrubs and hundreds of pounds of grass seed."[51] Such dedication makes us question our earlier stereotyping of Utahans as interested primarily in resource development, as exemplified by the building of dams or in their opposition to the establishment of national forest wilderness.

SALT FLAT

Salt Flat

West of Black Rock, U.S. 40, having lost its identity to Interstate 80, crosses two broad valleys and skirts the northern edges of two mountain ranges. Then it follows a straight line for forty miles across the Great Salt Lake Desert. Here, on top of a sheer rock point at the western edge of that desert and at the eastern side of the town of Wendover, Utah, the view is eastward, back across the salt flats that make the Great Salt Lake Desert famous.

As is suggested by the lack of major changes in the two photographs, the salt flats were, and continue to be, mostly an area to cross rather than a place to settle or develop. They have been, in fact, an obstacle to human travel for at least one hundred years. The main overland trail to Oregon and California went north around the desert, and the Pony Express route only brushed the southern fringe of it. The Donner Party crossed it directly, only to suffer later from early fall snows in California's Sierra Nevada mountains. In 1907, Western Pacific Railroad constructed the line visible in the photos, and in 1924 the then-unpaved highway was built "as close as possible to the track," Stewart commented, "since the railroad had been built first and materials were brought out by rail." The dependence on the railroad for construction of the modern freeway is not evident in the recent photo, where the dual lanes are almost invisible well to the north at the left side of the picture. The old highway still appears passable, at least for some distance, although most traffic from Wendover swings to the left toward an interchange at the interstate. "Billboards . . . on the left-hand side of the road, obviously placed to catch the eyes of west-bound motorists, and advertise the business firms of Wendover" are still present, although reduced in number. The American Guide Series book observed that in 1941, "where early trail breakers dared death from thirst and suffocation, dark-goggled tourists on two railroads travel in speed and relative comfort."[52] Today, the rails carry fewer passengers, but the freeway hurries even more people than ever, and at higher speeds.

Most tourists wish to cross these salt flats as fast as possible, but some enjoy driving out over them, even without bothering to travel to the Bonneville Race Track north of the photo view. Many tracks left by such "pleasure" trips are visible in both photos, a few the same; none suggests any hazard of getting stuck in the mud. The persistence of these tracks can be judged by the apparently good preservation of some of them for at least thirty years. It is even claimed that the route of the wagons of the Donner Party is still discernible. Another indication of the slowness of the land's recovery is the clear visibility in the 1980 photo of a small triangle of disturbed ground left isolated north of the pavement when the curve was rounded even before Stewart drove by.

It is unlikely that these salt flats will ever become recreational playgrounds, as another barrier to early westward travel, the mountain ranges, have become. The roadside rest on the interstate well out on the desert, with its tower from which to view the expanse of glaring white flatness, receives much use. But Horace Greeley's assessment over a century ago that "If Uncle Sam should ever sell that tract for one cent an acre, he will swindle the purchaser outrageously"[53] would probably still draw some agreement, even though Uncle Sam has in fact held onto the salt flats, using large acreages as bombing and gunnery ranges and briefly storing the very first atomic bombs just south of here.

Early pioneers who crossed these flats often were forced to discard their wagons and oxen, and continue on foot. The articles they left behind are frequently collected and displayed in museums. Future archaeologists will find fewer artifacts of contemporary travelers, except for shreds of tires that failed and the ubiquitous beer and soda cans. They may also wonder about the small stones collected from the freeway fill and arranged into letters and words on the salt beds in a unique sort of graffiti. Like an abandoned wagon, the solitary box car sits at the end of the railroad spur, undoubtedly not forgotten, but portraying by its starkness how out-of-place human activities seem in this landscape.

WENDOVER

Wendover

The character, if not the size, of Wendover, Utah, remains unchanged since Stewart photographed the town in 1950. The larger water tanks on the hill at the far edge of town suggest a greater need for water today than in 1950, although water continues to be imported by pipeline as it has since Wendover was originally settled. The trees in town have grown and are more numerous, but the settlement remains nearly as shadeless as the surrounding desert. Although signs announcing the presence of gas stations are considerably (and predictably) larger, at least two of the stations are in the same locations as in 1950. The buildings of Wendover continue to be "strung out along the pavement," suggesting the continued importance of highway traffic to the economic life of the town.

In fact, perhaps no other settlement along U.S. 40 in the intermountain West is as dependent upon highway travelers for economic activity. Other towns like Roosevelt, Utah, and Elko, Nevada, provide services for farmers, ranchers, and recreationists. But the area around Wendover is too poor in water for irrigated agriculture, too arid to produce forage for range livestock, and too lacking in relief to attract tourists. The town exists almost exclusively to serve transient auto and truck travelers. Of the twenty-seven retail businesses in Wendover, twelve are service stations for vehicles, six are motels, and six offer food. (Conspicuously absent among the food establishments are roadside taverns of the kind that are common outside Mormon Utah.) Stewart identified a similar array of businesses, but they were fewer in number. Wendover otherwise supports only one bank, one church (Latter-Day Saints), and the offices of several small industrial concerns and government agencies.

The great growth in highway traffic after World War II determined the town's character as a highway settlement. In 1945, the average daily number of vehicles passing through Wendover (excluding local traffic) was 497, and this number increased to 1,300 in 1950, when Stewart took his photograph. In the next twenty-two years, the traffic doubled, and features of the town reflect the greater activity. The photographs indicate that the town has gained not only additional permanent structures, but also a number of mobile homes, those symbols of rapid growth in many parts of the American West. The low cliff so prominent in the lower right foreground of Stewart's photograph has been nearly obliterated beneath two decades' accumulation of garbage and junk metal. As if to symbolize the reason for Wendover's existence, nearly two dozen old cars sit beside the dump immediately behind a line of gas stations and motels.

A new segment of freeway crosses the Utah-Nevada state line (made conspicuous by a change in paving material from asphalt to concrete), swings around a rocky outcrop on the right edge of the picture, and is lost to view. We see it again in the right background as it rises up the gentle slope leading out of the depression of the Great Salt Lake Desert. The old two-lane highway has become a frontage road; the slight jog in its alignment was made necessary to allow the new freeway to pass easily through the gap in the ridge.

It seems unlikely that the bypass will hurt the vehicle services of Wendover. The town is the only roadside settlement of any consequence in the nearly 200 miles between Wells, Nevada, and Salt Lake City, Utah. Travelers' needs for gasoline, food, and rest, particularly after or before crossing the barrier of the Great Salt Lake Desert, make a stop in Wendover a normal recourse, just as it was in Stewart's time.

PEQUOP SUMMIT

Pequop Summit

Highways that extend east–west across Nevada, like U.S. 40–Interstate 80, alternately climb over the north–south trending mountain ranges and sweep through the intervening valleys. Here the road is approaching, from the west, Pequop Summit, elevation 6,967 feet, in the mountains of the same name. Our 1980 photo has a wider view than Stewart's picture in order to include the newer highway ascending on the opposite slope. The pavement of the old road is still discernible below, although encroaching rabbitbrush is invading along cracks in the asphalt. The freeway seems to be rather old as well, and may be one of the first segments of the interstate to be completed in Nevada. Its age is suggested by the peeling paint of the guardrail, the heavy oil stains on the slow lane for eastbound traffic, a narrow dividing median, and the considerable growth of shrubs on the deep fill.

The highway is much changed, but like so many natural features along the western portion of U.S. 40, the vegetation appears about the same. The juniper tree in the foreground maintains its size and shape, except that the large limb on its right side has recently broken off; we found it lying browned but still identifiable in the brush below. Above the lower cliff face which rises to the left of the old road, many of the junipers and a few pinyon pines are still recognizable, and, in fact, have identical forms. The pinyon at the right side of Stewart's photo, which he described as "approaching more closely the form of a pine," is gone, having been removed along with considerable rock in the rebuilding of the highway. Stewart does not mention a recent fire in the landscape of his photo, but leafless and whitened snags suggest that a burn had swept up the slope beyond the foreground tree. Even in 1980, those snags remained as logs on the ground surface. This burn may have promoted the bunch grasses which Stewart said formed "nutritional pasturage . . . for either cattle or sheep . . . on such mountains as these." The fence which he noted "filling the gap not protected by the cliffs themselves," however, is gone, made unnecessary by the modern steel-stake fence lining the freeway right-of-way and extending across the old road on the curve below.

Like the vegetation, the rock cliffs and slopes have not changed over the last three decades. "The bold topography of the Pequops is typically far-western," Stewart wrote. The sparse vegetation and shallow soil of the arid West cannot camouflage the rock subsurface, or soften the steepness of slopes. In the humid East, by contrast, denser vegetation and deeper soil produce a more rounded and subdued topography.

It is not just the topography which distinguishes the landscapes of the East from those of the West. One critical factor which makes western terrain distinctive is the feature which more generally makes the West different from the East—dryness. The novelist Wallace Stegner comments that "there are many Wests . . . [but] they have certain things in common—aridity above all else, and the special clarity of light, the colors, the flora and fauna, and the human adjustments that have resulted from aridity." Another factor Stegner notes is the relative newness, the lateness of development, of the far West. In contrast, other parts of America yielded to "our need to 'break' the wilderness and plant our civilization." One consequence of this taming has been the emergence of what Stegner calls "nostalgic regret," an expression of "something quintessentially American: our sadness at what our civilization does to the natural, free, and beautiful, to the noble, the self-reliant, the brave."[54] But as long as parts of the West remain unbroken and uncivilized, hope remains that the benefits of wildness will persist along with the advantages of civilization. These sentiments may be accentuated at Pequop Summit not only by the "typically far-western" topography, but also by the presence of what is sometimes perceived to be the modern equivalent to the glorified cowboy of an earlier time—the "free, self-reliant, and brave" driver of an eighteen-wheeler bound for the pass.

SUN ON HIGHWAY

Sun on Highway

Thirty years have not dimmed the early morning sunlight that shines through clear, cloudless skies to provide brilliant contrast between the straight alignment of the road and the "rolling sagebrush-covered country" as it stretches eastward to the East Humboldt Range in the distance. As everywhere across Nevada, however, the highway, here halfway between Halleck and Elko, has become freeway.

Stewart mentions that this location is "a good example of the broad two-lane pavement" through Nevada, "especially widened with an additional shoulder" across the crest of a hill. Today's divided four lanes have been even further widened to accommodate more traffic at the high speeds that these straightaways seem to encourage. Newly laid and graded fill abuts the pavement in the foreground, thus providing a still broader shoulder for the crossing of the same hill. Despite the glistening surface of the freeway, our ride was far from smooth, because highway crews had roughened the asphalt in preparation for resurfacing.

The constant grading of shoulders along the state roads in Nevada, intended to allow for relatively smooth pull-offs in case of emergencies, limits invasive plants to small rabbitbrush and annuals like Russian thistle. Where the cuts extended beyond and above the shoulder, we occasionally saw crested wheatgrass, planted regularly by machine in rows. Such grass not only reduces windblown dust but also provides forage for range livestock.

In the early photo, the utility lines reaching far across the empty landscape seem to emphasize the distance between human settlements. In our photo, the telephone poles, their former individual lines now both cabled and neatly realigned with the freeway, as well as numerous power poles, are intercepted by a new spurt of development. To the north of the interstate, and reached from the Ryndon–Devil's Gate interchange at the foot of the grade, is a scattering of more than a dozen frame houses and trailers, with numerous small fenced pastures presumably for horses. This "sprawl" is representative of the areas of recreational home sites which have recently appeared near Elko. To the south of the highway is a luxury campground that could have been only a mirage a century ago when California-bound wagons may have spent a night on the flat. For a mere $8.50, a couple in a tent or "R.V." can relax on "level grassy sites" that feature a laundry, disposal station, grocery, lounge, recreational room, playground, and overnight horse boarding and corral. For another dollar, they can tap A.C. current for their color television, and in the winter 75¢ will rent a heater. Although a grassy pasture is visible to the right of the highway, heavy use of this valley by the horses as well as by cattle or sheep may be indicated by a relative lack of dense sagebrush and the predominance of disturbance-dependent rabbitbrush.

The scene may foreshadow the future, but it is not characteristic of today's Nevada in general. Marooned by a broken oil pan twenty or thirty miles north of here and forced to walk on a gravel road, contemporary author Donald Jackson described a landscape "between somewhere and somewhere else . . . silent as stone, empty as a ruined cathedral, as tough and uncompromising as the earth around it." Yet his attitude was far from negative. A "wilderness buff," he said, can be "damned glad that one can still get stranded on an American road."[55] We would guess that most of the patrons of Ryndon's Campground would not share his enthusiasm.

EMIGRANT PASS

Emigrant Pass

In 1950, Stewart found the "old road" at this site "wholly obscured" because of the cut and fill for two subsequent highways. In 1980, the wider road-cut for the fourth and latest highway forced us to stand to the right of the spot from which Stewart had taken his eastward-looking picture of the highway, sweeping in a series of curves past the tree-covered Emigrant Springs to the summit at Emigrant Pass. The modest two-lane highway of Stewart's day has become a divided interstate complete with metal median strip and guardrail.

Crossing the long but gentle pass, at 6,106 feet, is hardly like reaching "the top of the world" on the Continental Divide at Berthoud Pass. Stewart identified its location as "the mild sag in the skyline of the hills." In our photo, radio towers provide a new communication link across the pass. Motorists at the summit may also notice a turnout for truckers to check their brakes.

Four hundred feet in elevation below the summit was the real event for early passengers in covered wagons on the California Trail: Lower Emigrant Springs, visible near the overpass in the 1980 photo. There, according to the 1940 American Guide Series book on Nevada, the weary travelers "washed the alkali out of their throats and discussed probable troubles ahead."[56] At the time of that writing, the springs had been renamed "Primeaux" and had become a tourist and bus stop featuring local relics. In Stewart's time, service stations and overnight cabins still provided respite to the traveler, the springs being "the only water-supply along a considerable distance of highway." He noted that the native bulrushes, the main vegetation mentioned in early records, had been supplemented by planted cottonwoods.

Today the traveler cannot easily take the midday siesta sometimes recommended for travel through such hot, dry regions. A sign in our photograph reads "Emigrant, Exit 1 Mile, No Services." It is not that all modern travelers, thanks to the luxury of air-conditioned cars, escape the need for a siesta; on the contrary, the 1980 American Automobile Association Campbook still recommends driving only during the early morning and evening hours in desert country during the summer, and still acknowledges the psychological stresses of aridity and long stretches of unbroken roadway mentioned in pioneer journals (instead of "temporary insanity," however, it warns of "highway hypnosis," a breakdown of driver reaction at high speeds). The explanation for the decline of the stopping-place at Emigrant Springs may be a change in perception as to what constitutes "a considerable distance of highway." Emigrant Springs is less than an hour's drive from either Elko to the east or Battle Mountain to the west. Moreover, road atlases show a picnic site within ten miles of the pass, although when we stopped at the developed roadside rest we were greeted not by fine large cottonwoods but by small, withered saplings that evidently had died after a recent planting. Emigrant Springs does not even have a historical marker to entice visitors off the freeway. Much tall grass marks the area, but the "bulrushes" are gone. The oasis for travelers is now only a station for highway maintenance, and the area immediately beneath the trees is demarked with signs which say simply AREA CLOSED. Wildlife such as passing deer, pictured on one of the highway signs of the recent photo, are perhaps the only "emigrants" who enjoy the cool water and shelter of Emigrant Springs today.

TIE HOUSE

Tie House

Several of the dwellings in Stewart's earlier photos have disappeared from today's scene because of a variety of causes—explosion of a hotel in Atlantic City to make way for a gambling casino, dismantling of a stockade at Fort Necessity to allow more accurate historical reconstruction, demolition of the Baltimore rows for a high-rise and of Mount Prospect for a parking lot, and flooding of log cabins in Utah for a reservoir. None of these events were predicted by Stewart. But here, a few miles west of Valmy, he may have anticipated the demise by fire of a miner's house of railroad ties, saying that this threat was "a much greater danger" than either rain or snow, and noting that the stovepipe above the flat, tar-paper roof was both tall and capped "to prevent the escape of sparks." When we first glimpsed the cabin from the freeway in the mid-1970s, it was intact, but the charred ruins that greeted us in 1980 yielded no clue as to whether the destructive sparks spread from the stovepipe on the roof, from one of the two stoves in the remains, or even from vandals.

The area was deserted of human life as we wandered about waiting for the thick stratus clouds to clear enough to allow a photograph. No miner's dog, but rather a coiled rattlesnake beneath the crisscrossed collapsed ties, warned us from our investigation of the structure. The western kingbirds and meadowlarks flitting about were probably more long-time residents, as undoubtedly were the occasional jackrabbits. As was the miner, much of the wildlife is drawn to the waters of the small spring still flowing down the slope and to the resultant vegetation, which is more luxuriant than the surrounding sagebrush. Two of the four planted poplars that drew water from the spring were destroyed by the fire; one is standing, and the other is reduced to a blackened stump. The two pines at the left in the earlier photo, which Stewart observed to be less typical of the region and not thriving, have been replaced by native cottonwoods. The grass growing around the spring and previously protected from wayward cattle now has thickened and spread beyond the downed barbed wire fence and into the foreground, where it provides not only cover for wildlife but probably also additional feed for the three untended horses which grazed on the flat below us.

We find it difficult to believe that this site will be much changed in another thirty years. In this arid environment, the entry road, though lightly traveled, will probably be as devoid of invasive plants as it is today. The path to the mine up the slope behind us will also probably be as easily located, though its approach may be better barricaded for safety's sake. We see no recent attempts at mining activity, and the location is not near a present-day Nevada mining center. The remaining trees might survive to shade only the interiors of two wrecked automobiles, relics of the 1950s out of view of our photograph. The rusting water tank, added since Stewart stopped here, will also likely still be standing.

But elements of "progress" are on the horizon. Had the poplar to the left grown more slowly, the viewer would be able to see a smokestack much taller than that of the tie house. Located near one of the two railroads which parallel the highway to the north, and which may have supplied the ties for the miner's cabin, is the Valmy coal-powered thermoelectric plant. There is no danger here, however, of a strip mining controversy as in Craig, Colorado, for the coal used to generate the electricity is brought by rail from Utah to fossil fuel–poor Nevada. The water for cooling is provided by the Humboldt River. The power is then transmitted, in part at least, westward, as indicated by a line of steel towers and high-voltage wires far in the distance.

GOLCONDA SUMMIT

Golconda Summit

Stewart particularly liked places where U.S. 40 crossed summits or passes. He included a series of photos depicting the routes over Berthoud Pass and Rabbit Ears Pass. Later we will see a high-angle view over Donner Summit in the Sierra Nevada. In Nevada, he photographed three such places, this vista being the third. It looks westward from Golconda Summit over the valley of Rock Creek toward the Sonoma Range.

"U.S. 40 here swings to the left in a magnificent curve," Stewart wrote, "descending from the summit." The highway in 1980 was abandoned, but the freeway duplicates the "magnificent curve" immediately to the left of the photo view and at a higher elevation. The guardrails on the old road have been removed, as has the billboard whose presence in such remote country, Stewart said, should generate a "note of alarm." The pavement is in better shape than the abandoned road at Pequop; even the center line is still discernible. The fine drivable condition of old U.S. 40 along this stretch probably reflects the relatively recent date of its replacement by the interstate.

Stewart claimed to have seen evidence of three older roads at this spot, and old concrete culverts are still conspicuous. Moreover, he astutely anticipated the eventual replacement of the roadway in his photo: "Quite possibly the excellent highway here seen may itself become obsolete in another generation. If so, it will leave still another scar on the landscape for some future archeologist to decipher." We wonder if in the future the new freeway will similarly become "obsolete" and be replaced by still another road—or has design of conventional highways reached a kind of zenith, and will tomorrow's advances be toward new types of transportation systems?

Identification of the old road does not yet require the eyes of an archaeologist, even in the general sense in which Stewart used the word, but the encroaching rabbitbrush is beginning to reclaim the pavement. Stewart was correct in saying that shrubs like rabbitbrush are often confused with sagebrush in the "Sagebrush State" because "other desert shrubs give much the same general effect, and the tourist is likely to think that he sees sagebrush when he is actually looking at greasewood, or shadscale, or something else." Both photos reveal identifiable colonies of "something else": the fine-textured, oval-shaped patch on the slope to the left is winterfat, and the shrubs which in 1950 extended in a line just behind the guardrails and in 1980 covered the road fills are rabbitbrush. Away from the heavily disturbed soil beside the roadway, the vegetation is commonly perceived to be "sagebrush," although it is actually a mixture of different shrub species.

In the valley bottom, just beyond the dark band of greasewood along the course of Rock Creek, a whitish expanse appears in the recent photograph. It outlines still another patch of something other than sagebrush, a golden mass of annual grass which has colonized an area disturbed as a staging area for construction of the interstate. The line of dark cottonwoods beyond and to the right of the patch of grass is the town of Golconda, settled as a railroad town for shipping livestock. As with most other small towns along the freeway, Golconda seems to have suffered from travelers and truckers rushing by on their way to larger settlements. Earlier in this century it attracted tourists from the highway with its allegedly medicinal, and certainly comforting, hot-spring baths. Before the presence of the formal road, it was a camping spot on the California Trail. West of Golconda, according to an early account, "the light ashy dust once stirred went to rest only with the sun, and the air was filled with all sorts of prismatic hues; some of the travelers wore green goggles; and some wore bandages or little aprons over nose and mouth as protection against the dust."[57] Today, the modern traveler speeds by the off-ramp to Golconda, hurrying on to Winnemucca just beyond the Sonoma Range, with dark sunglasses rather than green goggles and with protective suntan lotions rather than bandages or aprons.

We enjoyed wandering about above the old road cut, along the abandoned pavement, and over the adjacent brushy slopes. A fine warm wind blew in the scent of sagebrush. In contrast to the usual wildlife along the highway—run-over rodents and scavenging ravens—here we enjoyed western kingbirds and lark sparrows, leopard lizards and jackrabbits. Raising our eyes to the snow-capped mountains farther south, we could easily agree with Stewart that the view, indeed the whole setting, was "typically Nevadan with a broad sagebrush-covered valley ending against rocky and jagged mountains."

IMLAY

Imlay

Stewart photographed several mountain summits and highway straightaways across Nevada, but he failed to portray even a single Nevada town. This rather distant view of the tiny settlement of Imlay is the closest equivalent to his earlier portraits of such places as Roosevelt, Utah, Idaho Springs, Colorado, and Hays, Kansas. Perhaps he felt that those earlier scenes adequately represented the streets of small western towns.

The far perspective of his photo is reflected in his written description of the community: "Having grown up as a railroad town, Imlay still turns its back on the highway. Its cottonwoods are a mark of Nevada towns, which can frequently be seen looming up as dark masses in the distance." We found Stewart's vantage point for viewing the settlement a bit too far removed. Exchange the high mountain peaks of the Humboldt Range on the horizon for tall grain elevators within the cottonwoods, substitute the shrubby greasewood flats with thick grassy plains, and the general form of pole lines, highway, and sky resembles the landscape far to the east in Kansas. We wanted a closer look at Imlay.

A drive towards town brought us to the lonely foundations of at least two abandoned roadside businesses near the turnoff from the freeway. In the rubble were still-legible signs reading *Little Nevada, Food,* and *Bar.* A few elms still survived in the concrete foundations, which were the sole remaining pieces of the old buildings. Having once "turned its back on the highway," Imlay's tourist businesses apparently were in turn largely ignored by travelers. The town itself does not seem to be prospering. We rode by modest homes and a small post office, but no stores of any kind. When we approached the railroad, the supposed mainstay of Imlay, we came to a good road, perhaps an even earlier highway than that in Stewart's photo, paralleling the rails. Again, however, numerous foundations lay gaping and abandoned, the remains of old railroad buildings, we assumed. Has the railroad turned its back on Imlay?

Stewart predicted that the highway in the future would "very likely" undergo the "greatest relocation" possible in its route and follow the shortcut from Battle Mountain to Lovelock, thereby bypassing Imlay. With even more certainty he foresaw that the pole line, "in all its magnificence," would be "pulled out and junked as soon as the telephone-company can get far enough ahead in its work to find the time" to replace it with buried cables along the same shortcut. Neither change has been realized.

The inertia of the past has been important in keeping Stewart's predictions from being fulfilled. The interstate follows the highway route of 1950, swinging "far off to the north, by way of Winnemucca, around the point of the range on the extreme left." The shortcut which Stewart anticipated would have bypassed Winnemucca, and thus would have deprived the town of tourist business and the traveler of needed services. The importance of highway traffic to the economies of the towns of Nevada is suggested by the fact that the very last segments of the freeway to be built have been the bypasses of the major towns of Wells, Elko, Battle Mountain, Lovelock, and Winnemucca. When we traveled across Nevada in 1980, the freeway ended at the outskirts of each of these settlements, the traffic moved down the main street, and the freeway began at the other side of town. It seemed as if the state was reluctant to discourage freeway travelers even slightly from stopping for a meal, gasoline, lodging, or perhaps a turn at the slot machines. The demise of the highway-service functions of Golconda and Imlay makes this seem reasonable. Such small settlements likely lacked the political influence to preserve even for a short time a more favorable situation. Winnemucca is too important a place to be sidestepped completely by a major "shortcut" of the highway. (We enjoyed, incidentally, the break in freeway travel that the slowed and varied passes through the towns provided, and wished that they might be maintained as permanent features of the crossing of Nevada.)

The persistence of the telephone line is more of a puzzle. We have seen the replacement of such a major pole system in the Kearney Hills of Kansas, and the disappearance of many minor lines such as those in the meadow at Rabbit Ears. Are service demands too great for the newer cables alone, or has the telephone company simply not yet "found the time" to "junk it"?

LAHONTAN STORY

Standing on the large, flat but gently sloping rock where we were sure Stewart had stood, we admired the great sweep of desert shimmering below us in the late-morning heat of early July. The view is eastward from a slope about ninety miles south of Imlay across the salt flats of the Forty-Mile Desert. This landscape closely fits the common image of the "Nevada desert"—barren, bleak, and blazing hot. Harry Truman's comments may resemble the feelings of many Americans who have not had the opportunity to travel on U.S. 40 across the "Sagebrush State":

> We flew over the remains of the Great Salt Lake, the saltiest of the salty seas, which is gradually drying up. Then we came to the great gambling and marriage destruction hell known as Nevada. To look at it from the air it is just that—hell on earth. There are tiny green specks on the landscape where dice, roulette, light-of-loves, crooked poker and gambling thugs survive. Such places should be abolished and so should Nevada. It never should have been made a State.[58]

We admire his style, which evokes such vivid images, even though we do not agree with his sentiments.

"In its name," Stewart commented, "the Forty-Mile Desert records the distance of the dreaded dry drive [for early emigrants on the California Trail] from the Humboldt Sink to the Truckee River." He implied that this segment of highway continued to be nearly as inhospitable in 1950, saying that it was "the longest stretch anywhere on the road to be completely without services. From Lovelock to the hot springs [in the Hot Springs Mountains], a distance of 41 miles, there are no supplies." In 1980, the landscape was generally as stark and lacking in the necessities for human life as in 1950, but the stretch with no supplies had been broken by the development of a store and gas station at the junction of U.S. 40 and U.S. 95, about midway between Lovelock and the hot springs. Still, with the loss of whatever facilities were available at the springs in 1950, there is still a thirty-five-mile length of highway between the junction of the U.S. routes and the outskirts of the town of Fernley without services. Thus, the Forty-Mile Desert and the Great Salt Lake Desert continue to rival one another for being the longest stretch of U.S. 40 that lacks supplies.

Both cultural and natural features display few changes over the last thirty years. The telephone line still marches across the foreground, the salt flat continues to shine "almost snow-white," and the lines of the old beaches of Lake Lahontan, "showing on the dark hill across the flat," are unchanged. The greasewood shrubs in the foreground show remarkable persistence, and many individual plants appear to have weathered thirty years with little change. For example, beginning with the bush near the base of the telephone pole on the right, a line of three shrubs can be seen extending toward the upper right and ending in a clump of several more. From there, a second line of seven shrubs angles back toward the lower left, the last two being the largest. Many other individual shrubs and clumps of shrubs are plainly identical in both pictures. The track of the old emigrant trail which Stewart identified "halfway between the poles and the highway, visible at the left," is no longer present. Heavy trampling by cattle, which gives the flat its spotted appearance, may have contributed to the obliteration of the old trail. The lines running back and forth among the scattered shrubs are not auto or wagon tracks, but cattle paths.

Livestock apparently find the scanty growth of grass, such as that forming the dark band along the near edge of the highway, highly desirable. We found many droppings of cattle along with hoofprints, but suspect that grazing was restricted to the spring months when the salt grass was more palatable. The grass at this spot persists because of the availability of water which seeps out from the base of the slope and backs up along the road fill, seemingly farther to the left in the more recent photo because of the additional fill required by the freeway. The water even forms several deep pools just to the right of the photo view. Apparently too salty for fish, the water and adjacent marshy lands do support other life. We observed algae thriving in the ponds, dragonflies darting overhead, avocets and killdeer wading in the wetlands, and both jackrabbits and cottontails scurrying away from our approach. The more common lizards and ravens, and the uncommon sweeping swallows overhead, added to the surprising richness of the place. Truman was wrong, we thought. Nevada is full of things that are good.

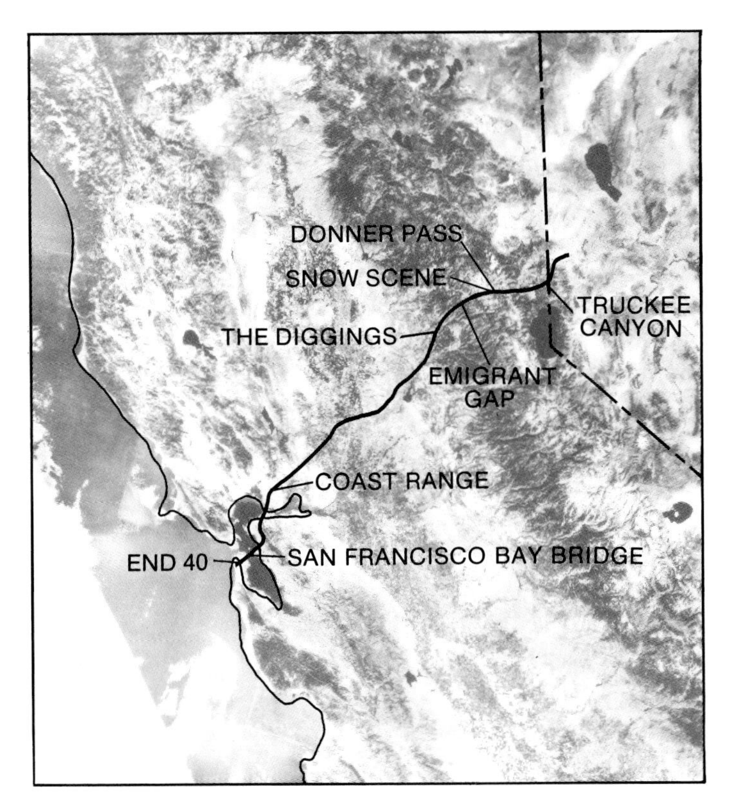

RENO TO SAN FRANCISCO: PACIFIC COAST

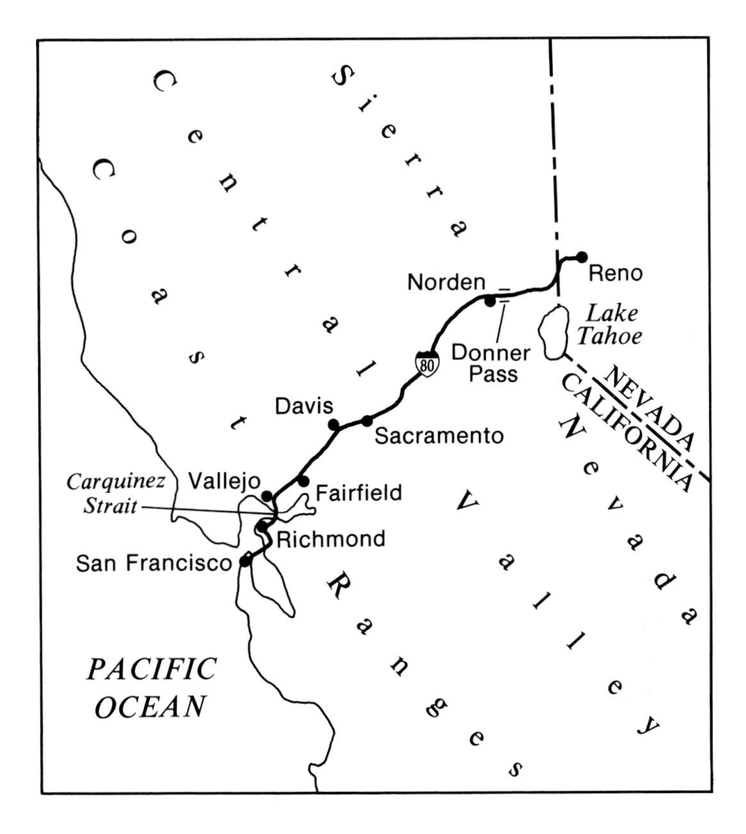

Reno to San Francisco: Pacific Coast

FROM RENO to San Francisco, the U.S. 40 of Stewart's day is followed today by Interstate 80. The only notable stretch where the highway has been relocated is over the summit of the Sierra Nevada; the freeway crosses the divide at a point two miles north of old Donner Summit. Nowhere, however, along the entire route across California does the U.S. 40 shield appear, because, as is true through Nevada, the Interstate 80 emblem has exclusively replaced it. Whatever its designation, however, the highway retains its historical significance of following as far west as Emigrant Gap the California Trail, by which overland wagons and walkers reached the settlements and gold diggings in the nineteenth century.

Stewart's selection of views on this segment of U.S. 40 is curious. His inclusion of eight pictures in the Sierra Nevada, with its varied and scenic topography and its long and rich history, is not surprising. But his neglect of lowland landscapes is conspicuous. He chose no scene in the Central Valley, and only one in the Coast Ranges; that single photo, moreover, emphasized highway construction rather than the setting. In his introduction to *U.S. 40,* Stewart disavowed any intention of using views simply because they portrayed something attractive: "In my pictures, I have tried not to enter into competition with either the 'esthetic' or the 'news-value' school of photography. . . . I have been interested in the typical as much as in the outstanding. I have sometimes even avoided the picturesque. . . . I reject nothing." Yet, his feelings about the lack of appealing landscapes in lowland California, particularly the Central Valley, were strong and may well have influenced his selection of photographs: "The Great Valley is probably as uninteresting as any district on the whole road, being as flat as the High Plains, yet lacking their illimitable sweep, and seldom displaying much effect of light or cloud. The Coast Range varies between great beauty in winter and spring . . . and dusty drabness in the late summer and fall." California west of the Sierra Nevada, however, included, as it does today, many features of interest: U.S. 40's crossing of the Sacramento River, the largest stream which the highway traversed west of the Mississippi; the great Yolo Bypass, a flood control channel three miles wide, filled with field crops, and crossed by a causeway in the heart of the "uninteresting" Central Valley; the fruit and orchard district between Davis and Fairfield, the only major area of orchards anywhere along U.S. 40; the bottleneck of transportation lines at the Carquinez Strait where highway, railroad, and shipping lanes converge in an area of high relief, a spectacular bridge, waterside industry, and hillside houses. Perhaps these features were too close to home and thus too familiar to Stewart for him to think of their inclusion, or maybe he was reluctant to add many photographs to this Reno to San Francisco collection, with its many views of the Sierra Nevada, but the omission of such interesting scenes in lowland California is striking.

The landscapes along old U.S. 40 through California have changed predictably over the last thirty years. Suburbanization has spread east from Sacramento to the foothills of the Sierra Nevada, both east and west from Fairfield in the Coast Ranges, and through the bayside hills north of Richmond. Service businesses for vehicles, travelers, and the masses of new residents have increased in nearly every roadside settlement below the higher elevations of the Sierra Nevada. The highway itself has been transformed into a multilane freeway, and only through the mountains is it restricted to four lanes; elsewhere, the highway broadens to six and even eight lanes. Traffic is much heavier, with a nearly constant stream of trucks, autos, and occasional buses, making this segment of old U.S. 40 the busiest long stretch anywhere from coast to coast.

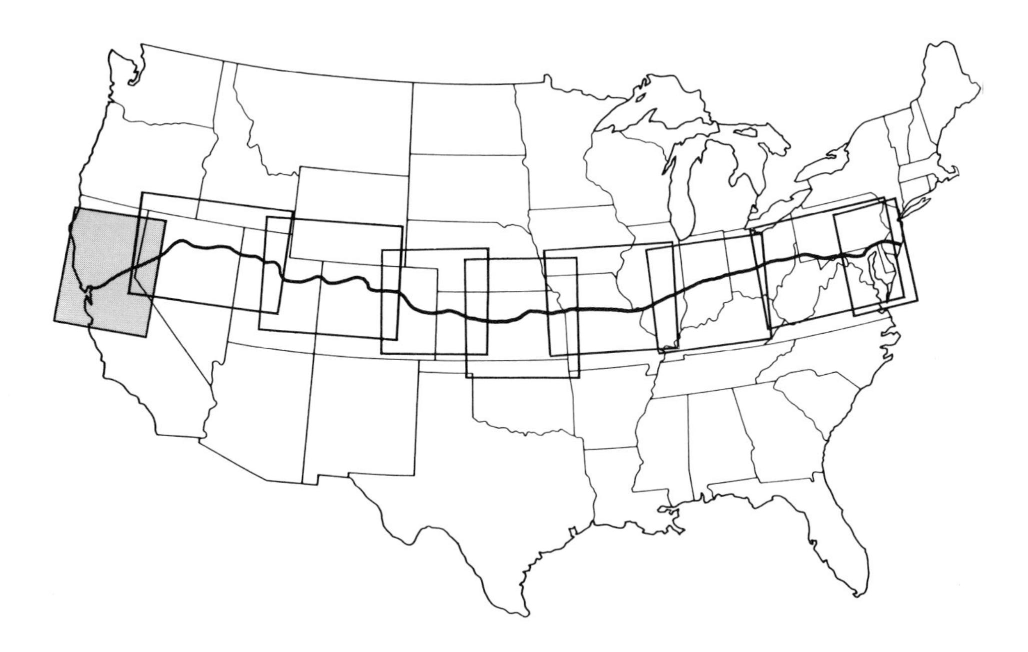

TRUCKEE CANYON

Truckee Canyon

Upon leaving the dry lands of Nevada, U.S. 40 climbs into the Sierra Nevada mountains, following the Truckee River. Here, the mountainous landscape today appears similar to that of 1950, but the highway itself has been conspicuously modernized to a full freeway.

Although a burn on the distant ridge has reduced a forested slope to a brush field, and the stand of trees to the left of the highway has thickened as individual trees have matured, the general character of the vegetation has changed little over thirty years. The openness of this pine-forested area, which Stewart ascribed to slow regrowth after logging and burning, persists today. The poor "recuperative powers" of plant growth in this relatively dry environment are especially evident in the rawness of the road cuts in both pictures. Even though in each view many years have passed since these slopes were created, they remain unvegetated. Desert vegetation, nevertheless, did not "continue to advance" as Stewart predicted. In fact, the pattern produced by expanses of brush and groves of trees remains remarkably constant. While preventing rapid expansion of the forest, the dry climate also apparently discourages its wide-scale destruction by eliminating the rapid accumulation of woody material and thereby discouraging large forest fires. This stability in the plant cover is in sharp contrast to the rapid revegetation we observed on disturbed slopes in the more humid eastern states.

The highway presents the most obvious change in the landscape, with four wide gleaming-white concrete lanes replacing the narrower dark asphalt roadway. Stewart's photo illustrates the perils created by impatient motorists on heavily traveled two-lane mountain highways, and certainly such perils must have been common on the Truckee Canyon highway of 1950. This section of road was upgraded to freeway standards in 1958 to accommodate the increasing traffic generated not only by vehicles on long-distance trips but also by those carrying California vacationers to and from the casinos of Reno. The right-of-way was widened by slight enlargement of the cuts and extension of modest fills into the river. One such fill in the foreground has displaced the flowing water to the right. The narrow four-foot dividing median reflects the relatively early date of the freeway design and construction. The concrete wall has been placed more recently within this dividing center strip.

In spite of its higher standard, the highway occupies the same alignment that it did in Stewart's time. This coincidence of the two roadways suggests the continued functioning of the canyon as a "communications bottleneck" in which highway, railroad, and power line find a passageway through the mountains beside the stream.

The canyon has funneled people, materials, and energy through the terrain, but no economic development has occurred within it. In part this absence of businesses reflects both the designation of this highway as a freeway, with the associated restrictions on roadside development, and the public ownership of most of the land. It also reflects the rough terrain devoid of heavy timber, rich mineral deposits, or abundances of other natural resources which might be utilized.

Still, the landscape is far from useless. Dams collect and store water for downstream irrigation and urban needs, brushy slopes provide forage for wintering mule deer, and the Truckee River offers sports fishing for both locals and tourists. The river also continues to produce electricity from small hydropower plants along its course, as the persistence of the power pole in the foreground illustrates.

In addition, the Truckee Canyon abounds with spots where a traveler may sit on a rock in the shade of a large pine and listen to the music of water rushing over boulders and sliding into pools. More than the rugged topography or the green forests, these welcome sounds tell the traveler that the long desert sector is over.

DONNER PASS

Donner Pass

When we talked with Stewart a year before his death, he expressed particular fondness for, and pride in, this fine view overlooking the final eastern ascent to Donner Pass. He might have equally admired the more recent photo, so little has this vista changed. The pockets of pine, juniper, and fir are still "growing in spots where they are sheltered from the prevailing westerly wind." Fields of shrubs, "looking from a distance like patches of moss," remain on exposed slopes where soil has accumulated. "The sheen of the superb Sierran granites" continues to give the landscape its distinctive Sierra Nevada look. The highway still follows as gentle a course as possible along the rugged slopes.

The freeway which replaced this stretch of highway runs through a gap beyond the rocky ridge that extends across the picture just beyond the bridge; a short segment is visible at the extreme right. In his little book on Donner Pass, published in 1964, Stewart saw this newest highway as the fourth stage in the history of roads across the crest: the mostly unimproved emigrant road, the formally constructed wagon road, the two-lane paved highway, and the four-lane freeway. "Whether the present highway will be the final word," Stewart speculated, "or whether indeed, it will even remain adequate for many years, must still remain doubtful."[59]

The old highway, no longer kept clear of winter snows, now serves as a scenic road for the summer tourist. The route is unmarked by signs or route numbers, and even the pass is not identified by name or even indicated as a summit. Thus, most visitors who discover it probably do so by accident. Two women who stopped at the observation area at the close end of the bridge while we were there, in fact, asked for directions to Lake Tahoe, and admitted that they had taken "a wrong turn" somewhere. They did not seem too upset by their mistake, because at least they had found "the top of the mountain." A family who also found their way to the parking lot seemed pleased that they were on "the old road just over the hill from the freeway."

People who stopped to take a picture or simply to admire the view of Donner Lake at the base of the grade to the right sometimes expressed curiosity about the history of the emigrant crossing of the pass. But no markers or interpretive signs provided help. In fact, unless they came with some knowledge about early overland travel to California, visitors could drive over the old road, stop at the overlook, and look at the vista without realizing that this was the route by which tens of thousands of people crossed the mountains with horses, wagons, and carts. Stewart expressed the importance of history to the personality of Donner Pass: "It has kept a rendezvous with history, and its interest to the person who passes here should be historical as much as scenic. At the summit, for instance, one can enjoy the beauty of the view, but can also see the remains of two primitive roads in addition to the present highway, can look across at the railroad, and can also know the emigrant wagons were dragged up somewhere to reach the same gap."[60]

We felt some concern about the future of the scenic qualities of this vista, specifically the fine arch bridge. Its concrete was weathering, and recent patching had been applied to the structural arch. If it should become unsafe, it may be replaced by a more modern steel overcrossing, or, worse (in our minds), by a fill of rock and earth. Certainly the ravine would require only a modest fill by modern standards.

The present bridge has a feature which reflects its construction at a time of less-hurried travelers: a little alcove and bench, which allow pedestrians to sit on the east side of the structure at the near end. We enjoyed a long stay there, watching the autos and trucks, greeting a couple of bicyclists who pedaled by, and admiring the long vista below us. Cliff swallows darted about, and a cool wind blew down from the pass.

SNOW SCENE

Snow Scene

A few miles west of Donner Pass and immediately west of the tiny settlement of Norden, old U.S. 40 crosses a heavily forested flat at an elevation of 6,900 feet. The more deserted look of the scene in the recent photo is not simply the result of the interstate having siphoned off the traffic. The obvious difference between the seasons is also critical. The highway near Donner Pass continues to be "a great road for winter sports," as Stewart commented, but that assessment understates the importance of snow to the local economy and the preoccupation with snow that characterizes the local population.

The heavy snows result in a winter-long battle to keep the highway open, and Stewart's photo suggests the magnitude of the problem because, as impressive as the banks of snow are, they represent only "moderate" conditions. "The snow-stakes are still standing up high, and have not had extra stakes attached to their tops." His picture also portrays a well-marked parking area, designated to concentrate parked cars in order to reduce problems for snowplows. Even though the 1980 road is no longer the main highway, it still is cleared of snow here, and the restrictions on parking seem unnecessary only to the summer visitor who looks over an apparently benign environment and a deserted roadway. Long "no parking" zones line the pavement, and frequent signs warn that illegally parked vehicles will be towed away. Such signs are typically attached to posts or structures high off the ground, lest they become buried by snow. Restrictions prohibiting parking altogether are posted on most houses beside the road.

Other features on this flat also suggest the deep packs of snow: fire hydrants with attachments for snow poles, and enclosed walkways which extend from front doors to the road shoulder. We saw, posted at a local store, a letter from the county director of public works answering a resident who had complained about inadequate snow removal; the official explained that plowing would continue to the same degree as in the past, and expressed concern over threats to crews and equipment. If harm were to be attempted, his letter continued, the state would no longer plow, and "should this happen, I do not know how the snow would be removed." The prospect of an end to snow plowing in this environment must be a sobering thought.

As much as the snow is a problem for highway crews, it is the reason most people come to this area. More than two dozen houses are located on this little flat, yet most seemed unoccupied when we took our summer photograph. The several ski areas, the nearly deserted lodges, and the boarded-up "ski clubs" are suggestions of the importance of snow to human use of the region. We were in the quiet little store at Norden for several minutes before anyone appeared behind the counter, giving us ample time to search the nearly empty shelves for a few supplies. "We do most of our business in winter," was the clear explanation offered by the young clerk, when our investigation proved unsuccessful.

Stewart concluded his discussion of "Snow Scene" by noting that "U.S. 40 was blocked by heavy snows for four weeks in January–February, 1952." When the segment of freeway over the crest of the Sierra Nevada, which bypassed the flat in the photograph, was completed in 1960, one of the proclaimed advantages of the new road was that winter closures across the summit were events of the past. In its first season, however, the freeway was forced to close by the heavy snow and high winds of a Sierra blizzard.

EMIGRANT GAP

Emigrant Gap

Here, as at Donner Pass, Stewart was again interested in history as much as in scenery. In his discussion of Donner Pass, Stewart lamented that the highway builders had obscured the traces of old trails so completely that despite frequent scramblings among the granite boulders over twenty years he could conclude with certainty only that "somewhere" along the pass early travelers painstakingly made their way. In his search of this site, he was again frustrated in his attempt to reconstruct the past. Early emigrants, he knew, originally brought their wagons through Bear Valley, to the left of the photo view, but soon established a better route across Carpenter's Flat, in the upper right of the picture, and through a gap in the watershed divide now occupied by both railroad and highway. Stewart noted that the location of the crossing of the ridge was doubtful because of both railway and highway construction work. He guessed that the low spot was in 1950 either buried beneath the railroad fill in the distance or paved over with the highway under the overpass closer to the foreground. A historian's job today would be even more difficult; our photo had to be taken atop tons of more recent fill.

This newest earth reshaping and consequent history erasing could not have been a surprise to Stewart, for he wrote in 1950 that, with heavy traffic and a blind curve, "the highway cries out for relocation." Like the earlier guides, modern engineers have achieved a "better route." The new freeway still follows the ridge, but now on its raised bed passes over, rather than under, the railway, and the driver's view is unrestricted. Four lanes accommodate heavy traffic, and the divided highway on the slope in the background allows a more gradual upgrade. The utility lines disappear underground at the freeway but emerge on the other side. The old highway shows as a frontage road near the meadow in Carpenter's Flat to the right, more visible than in Stewart's photo.

The pull-out in which we parked apparently also differed much from the one in Stewart's day, which was bordered neatly by white wooden posts but lacked curbing and an access lane. Today's pull-out, although reachable by westbound vehicles only, is well marked as a "Vista Point." The few tourists who pulled into the overly large, practically empty parking lot showed no difficulty finding the "vista," but appeared confused about the location of the old emigrant trail and gap. Although a Forest Service map identifies the topographic features and a historical marker recounts the story of Emigrant Gap, the conversations among those who studied them revealed they were little more informed than the tourists at Donner Pass, who had to glean all their information from other passersby. One reason for their confusion was that apparently, when the pull-out was modified, the historical marker had been moved from its original location and realigned not with the landmarks it described but with the orientation of the new retaining wall.

We suspected that the people who stop in the 1980s are not too dissimilar from those who might have stopped thirty years ago. While we were there, the spot was visited by one family with young children, several older couples stretching their legs, two young men with car trouble, a single man who took a short stroll while smoking his cigarette, and a young couple who enjoyed each other's company as well as the view. A sign in plain sight read $500 FINE FOR LITTERING, but the scattered litter testified to the persistence of a population's habits. The rock in Stewart's picture above the highway with its graffito, "Christ Died for Our Sins," probably disappeared along with the old road, but new carvings are on the informational signs and the few nearby trees with trunks large enough for a scratched initial or two. However, we decided, the loud music that emanated from the open doors of the young couple's pickup and carried to their seated location thirty feet distant was definitely a characteristic of the 1980s.

An excursion around the area confirmed our earlier suspicion that most travelers probably postpone their short stops for the rest area fourteen miles ahead. An older lodge on the flat by the old road served meals in a pleasant setting and atmosphere, but during limited hours only, and its many motel rooms, although they looked attractive from the outside, seemed to be vacant. The settlement of Emigrant Gap, west of the interchange along another frontage road (a side road even in Stewart's day), was even more deserted. Former stores appeared long-abandoned, as did all but a few of the residences. A road sign warned of a school, but we could not locate it. The great freeway fill rose steeply behind the town, obliterating the famous spot which gave it its name but not quite burying the buildings themselves.

THE DIGGINGS

The Diggings

Most people who drive along the "high red cliffs of gravel" about thirteen miles west of Emigrant Gap may not realize that this attractive escarpment is the product of hydraulic mining, an activity that would be decried as environmentally destructive today. The American Guide Series description of this area only hints at the massive, human-created erosion responsible for the landscape, and the image created by the passage is strongly positive: "The highway cuts into a deep gash along the face of bluffs tinged with vivid hues of russet, buff, and rose red. Over the canyon bottoms sprawl mounds of reddish earth and gravel, half overgrown with scrub pine; beyond rise crumbling pinnacles and blasted palisades."[61] The great hoses which were trained on these cliffs and which washed gravel into sluice boxes where gold could be extracted have long been outlawed in California because they caused flooding and sedimentation on valuable farmland downstream. But the spectacular erosional forms they produced have endured, not only along the highway here but also in an area farther north where a state park now protects the scenery.

Stewart described the forest at the top of the cliff, "where the topsoil was never washed away," as "ordinary second- or third-growth Sierra forest," but below the highway he said that the vegetation was "devastated." He predicted, however, that "in another thirty years the two sides of the highway may be superficially indistinguishable." But thirty years seems to have been too little time to allow the recovery of the forest to the left of the highway. It remains an open stand of small trees with much brush and bare rocky soil; the slowness of vegetative development attests to the impoverished, droughty substrate created by the mining. In contrast, the trees at the base of the cliff at the distant curve have grown well. The right foreground in the recent photo is not represented in Stewart's view, but the handsome pines there are large, and their seed has probably contributed to the establishment of the young trees along the edge of the highway.

The freeway was constructed along the same alignment as the old road, with the wider right-of-way created not by cutting back the cliff, but by extending the fill into the area left of the pavement. The westbound lanes adhere closely to the highway in Stewart's photo, although the concrete pavement and undoubtedly the rock base for the pavement are new. The heavy traffic prevented our running out to the edge of the narrow dividing median in order to duplicate Stewart's view precisely. Instead, we walked out an on-ramp, stood close to the rushing cars and trucks, and approximated his view. We can only guess about the origin of the tire marks which are visible on the access lane. The dark set of marks on the shoulder probably was left by a vehicle or vehicles which had inadvertently crossed an area of fresh tar. The other mark, much fainter, which extends from the lower left corner and crosses the ramp lane toward the upper right, likely was caused by an auto, perhaps skidding off the freeway as it unexpectedly came up upon a slow truck just entering the right lane.

At the far end of the cliff visible in the recent photo, a roadside rest allows travelers not only to rest but also to learn about the hydraulic mining which created the escarpment below which they have just driven. This "ecological viewing area" includes old machinery, photos, and a self-guided interpretive trail which extends away from the picnic tables and parking lot. We wandered about the exhibits for an hour but encountered no other people. Most people who stop at the rest area apparently use the rest rooms, eat a meal, or sit in the shade, and then drive off without learning that the "high red cliffs of gravel" were, at least in part, created by people.

COAST RANGE

Coast Range

Probably no stretch of U.S. 40 from coast to coast, and certainly none photographed by Stewart, has been widened and otherwise "improved" more than this segment in California's Coast Range between Fairfield and Vallejo. An earlier stage of reconstruction of the old two-lane highway was shown in Stewart's picture: "In the fall of 1949 this section of U.S. 40 . . . was being transformed into a four-lane freeway, fully modern in all respects." By 1980 the "fully modern" design was only a memory, having been replaced by a grand eight-lane, split-level freeway within an immense right-of-way. Stewart's comment about the impact of the then-unfinished highway seems an understatement for the 1980 road: "The gigantic fill has changed the whole appearance of the country, the opposing hill has been cut away for rock and gravel. What was apparently the bed of a small stream has been completely made over, and the stream has been diverted through a large culvert and into the drainage ditch—emphasized by the shadow—on the side of the highway." The "gigantic fill" is today unrecognizable, having been itself bordered by still greater fills. The cut on the opposite hill has been enlarged slightly, but it seems a minor modification of the topography compared to the bench produced by cuts and fills which permit the gradual, even grade of the westbound lanes. The stream which had been transformed into a drainage ditch is now buried entirely as an underground culvert; the grate in the foreground leads down into the subterranean channel.

In contrast to the highway, the clump of live oaks on the hill is so unchanged that individual trees are recognizable in both pictures. "Just why these trees grew in such clumps," Stewart wrote, "is a question difficult to answer surely." The great stability of this grove over the last thirty years suggests that one answer Stewart offered, frequent burning, is inadequate by itself, because fires are likely too uncommon today to prevent expansion of the trees into the grassland. On the other hand, the blackened grass and small shrubs in the foreground indicate that fires begin and spread readily in this summer-dry climate. Probably more influential, though, is livestock grazing, the other factor Stewart suggested. This environment "of muted tones" is "excellent grazing country" today, as it was in Stewart's day. The persistence of the "closely spaced parallel paths beaten down by cattle, faintly observable at the right of the picture," indicates its continued heavy use.

The emptiness of the landscape is deceptive. Just beyond the distant hills lie the orchards and rapidly growing subdivisions of Fairfield, and a few miles to the west are the housing developments and urban centers around San Francisco Bay. The heavy traffic and high-voltage power lines both hint of these large nearby human populations.

The two vehicles off the pavement in the early photo were discussed by Stewart. The one on the left was a "rugged, or extreme, individualist" who was trespassing on the unfinished roadway in order to pass the long line of cars trapped behind the trucks, and the one on the right belonged "to the photographer, who crossed the pavement, climbed the fence, and went a little way up the hill for his picture." Like Stewart, we trespassed on the adjacent grazing land, climbed the fence, and went up a hill, but then were forced to angle carefully back and forth down the steep new freeway cut before taking our photo.

SAN FRANCISCO BAY BRIDGE

San Francisco Bay Bridge

This is the road that no man finishes traveling . . .
This is the bridge that no man crosses wholly—
lucky is he who through the mists and rain-clouds
sees, or even believes he dimly sees, the farther shore.

When Stewart wrote these words about the Bay Bridge in his fictional *Earth Abides,* he meant them figuratively. In the novel, the bridge, which he describes as "the largest and boldest work of man in the whole area," is a symbol of the unity and security of civilization. During our attempts to duplicate the view of the skyline of San Francisco, we encountered some unique and particularly troublesome problems, which resulted in as many as seven re-crossings of the western span in a single early morning, and made us consider a more literal interpretation of the quotation. In *U.S. 40,* Stewart reported that he took his photo on a Sunday morning from a slowly moving convertible. Both pedestrians and stopped cars are still prohibited on the bridge, so we had to replicate Stewart's procedure of taking the picture from a car in motion. Even early on Sunday morning, though, when we expected the traffic to be light, we found ourselves in a nearly constant stream of autos and trucks. The rapidly moving vehicles not only blocked our view of the city but also made slow driving hazardous. We were thus forced to cross the western half of the bridge repeatedly, each time speeding up to get ahead of clusters of cars, and then slowing down long enough to snap the shutter. Each trip also required that we turn around in front of the stern military guards at the entrance to Treasure Island, midway on the transbay crossing, thereby making our trips less than "whole." In comparison to these difficulties, the high fog of summer, which on two mornings was too thick for taking photographs (we were not lucky enough to see "through the mists"), seemed a minor inconvenience. Our resulting photo may not be from the same point as Stewart's. The view, however, encompasses the same features, so that we can make comparisons.

In *Earth Abides,* Stewart makes clear his opinion that the bridge would resist deterioration through time even without maintenance. This feeling is expressed as the viewpoint of the main character who gazes at the unused bridge after half a century. Most of civilization has been destroyed by a virus: ". . . he saw the tall towers and the great cables, still dipping in perfect curves. This part of the bridge seemed to be in a good state of preservation. It would apparently stand for a long time still, perhaps during the lives of many generations of men."[62]

In the real world, three decades of constant use have been paralleled by thirty years of upkeep and remodeling. The number of autos, trucks, and buses which cross the bridge daily has increased over the last three decades, from 80,000 in 1950 to 100,000 today. The pavement has been changed to accommodate the greater number of vehicles. In Stewart's day, six lanes for autos, three in each direction, were available on the upper deck, and three lanes for trucks and buses (the central lane for passing), plus two tracks of an electric train, were on the lower deck. Today, the train is gone, and the decks each have five lanes, westbound on the upper and eastbound on the lower; autos and heavy vehicles are no longer separated. Beyond the lanes and the narrow bordering footwalk for workmen, the early photo reveals a sturdy guardrail above the protective solid steel wall. Stewart regretted this as "one of the few blemishes in the architecturally magnificent whole, for to an unnecessary degree the parapet and railing produce an effect of heaviness and interfere with the view." By 1980, at least the railing had been removed, and the vista was noticeably improved. The old lamp posts have been replaced by more modern streetlight standards. The "great cables" supporting the roadway are, predictably, unchanged, and "still dip in perfect curves."

In contrast to the modestly changed bridge beneath the cables, the skyline beyond the cables is so completely altered that two different cities seem portrayed. Over the last thirty years, San Francisco has declined as a port, but it has grown greatly as a corporate and financial center, so much so that the downtown today is sometimes described as "the Wall Street of the West." These functions seem to require the large number of offices in close proximity made possible by the modern skyscrapers. The decidedly contemporary personality of the scene does not seem to detract from its attractiveness. As in Stewart's day, we agree that "no city along U.S. 40 equals hill-built San Francisco in the bold picturesqueness of its skyline." We think, too, that its natural setting on bay and coast and its streams of summer fog are as important to its character as its concrete-covered hills.

SHEL
SHELL
PAPERS HERE
NEWSBOYS
PULL to CURB
M. A. GERAGHTY Co
DEALERS IN
Metals Scrap Iron
1374
END
40
50
SALVAGE MATERIALS
BOUGHT SOLD
PLATING

END 40

CHROME
Have a CC summer
TEXACO
TUNE-UP BRAKE REPAIRS
380
FRA TRUCK

End 40

The corner of Harrison and Tenth in downtown San Francisco is no longer the end of U.S. 40, nor of U.S. 50, nor is it an intersection along U.S. 101, nor does it have any other particular significance for highway routes. Although the junkyard and one of the two service stations which Stewart described as characteristic of U.S. 40 are no longer present, the neighborhood is still dominated by highway vehicle businesses. The Shell station is gone; its replacement, a self-service car wash, provides a vehicle service apparently unavailable here in Stewart's day. Down Tenth Street to the right, Frank Stuber's truck shop still occupies a large portion of the middle of the block. Farther down Tenth are businesses offering wheel service, brake maintenance, and body work. Beyond the Texaco station on Harrison, the building with the word *Chrome* on its side is much enlarged over that of 1950, but the type of service seems unchanged; a smaller sign in the early photo proclaims, "Plating—auto and household ware."

Quite apart from the neighborhood's continued function as a service area for cars and trucks, its personality seems much the same as thirty years ago. The wires of "the trackless trolley" are still "carefully suspended" above the pavement, even though one of the supporting metal posts has been moved to the right. The streetlight standard at the corner is unchanged, although the glass globe is slightly more rounded today. The billboard at the extreme right continues "to turn its back to the photographer," but a new billboard beyond the service station stares straight back at the camera. The large brick building behind the chrome-plating shop is an old school, likely functioning as such in 1950, but no longer in 1980, probably one of the victims of declining student numbers. Its rooms, once filled by children of the postwar baby boom, are now used exclusively by an older clientele for audio-visual education. A distinctly contemporary message, and one entirely appropriate for San Francisco, warns persons today who enter the rather shabby courtyard: "This building does not meet earthquake safety standards." The street itself is as deserted as it was in Stewart's view, but the lack of vehicles is just as deceptive. Both photos were taken on Sunday mornings, which allowed the 1980 photographer to follow Stewart's example and "stand in the center of the street without endangering his life."

The weekday traffic is still heavy at this corner, but the intersection has lost its significance as the end of a great transcontinental highway. U.S. 40 today disappears into Interstate 80 more than 750 miles to the east, in the Wasatch Mountains in Utah, and thus has as informal a western terminus as its eastern one in Atlantic City.

Perhaps the interstate routes will with time become the sentimental favorites of writers of books and songs, as were such roads as U.S. 40 and U.S. 66. But because the interstates isolate the travelers on the highways from "the land that lies beside [them] . . . and the streams that flow beneath [their] bridges," it is also possible that they will never find an abiding place in the human heart.

We are saddened by the loss of the special character of old U.S. 40. Our photos reveal that much of the pavement of U.S. 40 remains drivable and that most of the features which Stewart portrayed are still identifiable, but that the essence of the highway, the reality that was created by a shield with a number on it, will never be again. In 1950, Stewart concluded his epic narrative with the comforting reflection that the spatial end of this continent-spanning highway at an obscure city intersection was appropriate for a "working road." In 1980, the "job" of this once great highway, though not of the route it follows, seems perhaps "finished," in time as well as in space. Stewart's words take on a double meaning for us as U.S. 40 approaches a much more final "ending"—a lapse into obscurity in a still modernizing nation perhaps pursuing "its own affairs" with a speed and efficiency not attuned to the "romance" of the past:

> U.S. 40 is no swank boulevard, no plush parkway. If there is romance about it, this is the romance of the modern world, busy with its own affairs. U.S. 40 is the thoroughfare of a hundred thousand trucks and buses, and of a million undistinguished coupes and sedans, convertibles and station-wagons—driven east, driven west—for business or for pleasure, twenty-four hours a day, in a seven-day week. Most of all, U.S. 40 is a working road, and when it dead-ends into U.S. 101, its job is finished. Why should we seek an artificial climax? Work done, why should a road—or a man—make heroic gestures? End—U.S. 40.

Reflections

OUR TWO TRIPS on U.S. 40, the first vicariously with Stewart in 1950 and the second by ourselves in 1980, allow us to compare not only the changes in particular scenes but also the changes in the American landscape more generally over the past thirty years. Reflecting on the photographs as a group rather than individually brings several thoughts to mind.

Before looking through these pages, a reader might have anticipated sweeping transformations along U.S. 40. After all, the thirty-year period from 1950 to 1980 was a time of economic and population growth in the United States unprecedented in recent history. Changes in the American scene to parallel that growth would be a logical expectation. With a similar expectation, we began our journey in 1980.

We looked for the landscape changes about which Americans are so accustomed to hearing—the ticky-tacky houses, the unsightly commercial strips, the paved-over farmlands, the ugly freeways. We did find many places where buildings had replaced farms or pastures, and where multilane super highways had superseded small two-lane roads. Moreover, such places were widely dispersed, as suggested by the new suburbs in the Middletown Valley of Maryland, by the commercial and residential development on the slopes rising from Denver, and by the new freeways on Little Savage Mountain, on the Great Plains near Limon, and through the Pacific Coast ranges. But we also drove along many streets lined with well-kept houses dating from the days of early urbanization, by countless farmhouses remodeled only slightly since first settlement, through vast acreages of wheat, corn, and alfalfa, and over hundreds of miles of leisurely paced roadways. Indeed, although a few views presented strikingly different personalities from those of the past, more modest changes characterized most of them. For every Mount Prospect, Kearney Hills, or San Francisco skyline, we found more than one Atlantic City, Page City, or Imlay. Perhaps, then, the most surprising thing that the photo sites revealed to us was the lack of great change.

Some readers may suggest that U.S. 40 underrepresents change in the American scene because it has become a minor road, removed from the mainstream of progress. We disagree. First of all, the route of U.S. 40 is far from being a minor one.

Moreover, consider a contrary extreme—suppose that the entire length of U.S. 40's modest pavement of 1950 had become one-half of a spreading interstate freeway by 1980, its right-of-way marked by the dramatic changes associated with freeway development. Our photographs taken along such a U.S. 40 would illustrate a highway everywhere modified over the last thirty years, but we would be clearly wrong if we interpreted the photos as suggesting that all highways, or even all major highways, in America had become freeways. Most miles of most routes in the United States are not freeway today, and we suspect that they remain, like the landscapes through which they pass, much as they were in 1950. Secondly, because most of us live in or near urban areas that have grown considerably over the last thirty years, we tend to think of such growth as typical for the entire country. The United States is so big, however, so full of small towns, farms, pastures, and empty lands, that the great growth around the edges of large settlements and cities and the transformations of the rights-of-way of many major highways are actually localized phenomena. Watchers of landscapes may with some reason decry the spread of human settlement and the upgrading of highways, but such developments have hardly engulfed every rural acre or every winding mile.

The photographs in this volume, if anything, overrepresent the transformation of the American road from a two-lane highway into a multilane freeway. A truly random sampling of seventy-two sites around the country would reveal many fewer incursions of freeway than do our photographs. The route of U.S. 40 remains today a highway corridor more major than many other continuous 3,000 miles of 1950 roadway. But in addition to showing us the emergence of the interstate freeways, the photos also illustrate that the American roadway has experienced a varied history since 1950: in places it is now abandoned, in places a minor road, in places a major two-lane highway, in places a four-lane highway, and in places a multilane freeway. The photos provide a cross section of changes in the American roadway over the last three decades.

Another way of using Stewart's photos as a basis for documenting landscape change would be to travel along a transcontinental interstate and photograph scenes that portray the modern equivalents of Stewart's themes. Such an exercise might produce photos of a new freeway passing through downtown Baltimore as a counterpart to Stewart's "Baltimore Rows," a wide, prestressed concrete bridge across the Mississippi to match with his picture of the narrow, steel-arched Municipal Bridge in St. Louis, and a broad freeway with massive cuts and fills rising gradually up a mountain slope as the contemporary equivalent of Stewart's views of the winding roads over Berthoud and Donner Passes. Such a project would have merit because it would illustrate how a major highway of 1950 and the landscapes beside it compare with a major highway and its settings of 1980. It would, however, emphasize change by looking for it. It would not give the more balanced, and we feel more honest, look at landscape change that the photo pairs in this book provide.

Besides the common belief that much of our nation is wracked with change, another popular contention is that these changes have contributed to a pervasive sameness in the landscape. Again contradicting expectations, the photo pairs seem not to support the notion that the expansion of national retail chain stores, or the spread of certain architectural designs and materials, or the increase in farm size and technology has homogenized the American landsape to the point that Kansas looks like Maryland, or Colorado like Ohio. Admittedly, we noted that lines of dark trees in nonforested landscapes often mark distant towns whether in Kansas or Nevada, and we remarked upon the similar appearance of farmlands along the Atlantic Coast and in the Middle West. But these similarities have little to do with recent standardization in business, architecture, or agriculture. Even in 1950 Stewart could say that "a picture of one city stands pretty well for a picture of another," and, whether in 1950 or 1980, a close-at-hand view of a corn field would look much the same in New Jersey, Illinois, or California. Moreover, although some contemporary landscape features, especially those of human development, may seem more uniform to us today than those which had been established by 1950, perhaps the cultural artifacts of any particular vintage have much similarity. In Stewart's day, for example, the proliferation of elm-lined streets in small-town America may have appeared at times as monotonous to some travelers as do the stretches of fast food chains of the suburbs to a modern traveler. It is more than conceivable that some unforeseen economic or social force will continue the evolution of roadside eateries beyond the Burger Kings of the 1980s, as Dutch Elm disease has influenced the pattern of street trees since the 1950s. Ornamental plantings, architectural styles, highway designs, and economic activities often vary spatially *because* they have varied temporally. The coexistence of older, old, and new provides anything but dull uniformity.

We interpret the photos as suggesting, then, that landscapes have not been transformed dramatically everywhere, with commercial development rapidly engulfing the rural scene, and with standardization obliterating the unique. Instead, what has characterized landscape change is quite the reverse. The American landscape remains, as it was in Stewart's day, a study in diversity. The expansive urban centers on the Atlantic seaboard, the patchwork of pastures and woods in the Appalachians, the innumerable fields of corn and wheat in the nation's midsection, the rugged mountains and empty deserts in the interior West, and the varied scenes on the Pacific Coast; old stagnant cities and vibrant growing cities, small towns and spreading suburbs, coal mines and oil wells, cattle and sheep, grasslands and forests—all still greet the traveler along U.S. 40. A trip along the route provides the traveler today, as it did in Stewart's day, with a balanced impression of the United States.

One type of landscape along the route that particularly intrigued us is what some have termed a kind of "middle" landscape, one neither producing products for people nor formally protected for its natural features. Documented by our photos, it typifies roadsides throughout America. Wild nature persists in such places, often as spots where young children build "forts," where passersby admire the new leaves in spring, or where explorers both young and old search for a glimpse of a chattering squirrel, for the relief of a bit of shade, or for the contentment found in nothing more particular than a brief respite from the "madding crowd." This wild nature is the wild nature of the everyday world, persisting in places usually neglected by farmers and business people, overlooked by park administrators, but nearly all used by someone in one way or another. Middle landscapes include creek bottoms in Funkstown and Truckee Canyon, wooded slopes rising beside the highway at the Mason-Dixon Line and at Berthoud Pass, isolated groves of trees in the Kearney Hills and in the Pacific Coast ranges, roadside ponds in the Nevada desert and in the Wasatch mountains, and even bushy backyards "worth finding out about," as E. B. White described them,[63] in Harmony and Heber City. They help to bring nature into our everyday lives—into the long periods between planned vacations.

If we dare generalize at the subcontinental scale, these middle landscapes seem scattered from coast to coast. Productive landscapes, on the other hand, seem characteristic of the humid climates and gentle terrain in the eastern half of the country, whereas the potential for protected landscapes is greater in the drier and more rugged West. Still, our photos encompass surprisingly little landscape that is officially protected. The Colorado hogback is within formally established open space, but otherwise the great parks and wilderness of the American West are always in the distance—beyond the background ridge at the Utah-Colorado state line, and far off to the north at Donner Pass. Even so, expanses of wild nature that seem just that—wide and lonely—persist along U.S. 40 even without formal protection in the Great Plains, the Rocky Mountains, the Sierra Nevada, and especially in the sagebrush slopes and salt flats of Nevada. The wild character of the West owes its personality more to these unprotected lands, extending back from the pavement, that support a few livestock, a little logging, a scattering of alfalfa, or an occasional mine, than to the few and remote patches of more "pure" wild nature in parks. The West without the Dinosaur National Monument or the Desolation Wilderness would be, although poorer, at least conceivable. The West without winding roads ascending steep mountain slopes, impatient logging trucks barreling across flat valley stretches, cafes or gas stations dotting the clearings, and a ski resort or a campground looming in the distance, or without roads stretching across empty valleys of sagebrush, cattle grazing indifferently along the shoulder, and power lines disappearing at the horizon, would not be the West at all.

At this point, even the casual reader may begin to question the consistency of our attitude. At times we admired new freeways, discerned a pleasing harmony in the street patterns of a housing development, and even enjoyed interpreting the features of a commercial strip as a portrayal of changing forces and values in our economy and society. At other times—certainly at Mount Prospect and Starvation Reservoir—we openly lamented the sacrifices made for the sake of new developments. The two conflicting positions reflect our ambivalence toward landscape change. Two personal experiences, one of the first-named author of this volume and the other of Wallace Stegner, will help illustrate the dilemma that we see.

Recently on a clear December afternoon, my parents and I walked on a newly established recreation trail through the hills at the edge of a suburban town east of San Francisco. The winter rains had been abundant. The slopes were lush with the dark green of knee-high grass, although the scattered deciduous oaks still withheld their leaves, awaiting the warmth of spring. A few mustards were in bloom, but the cold wind that bowed their golden heads felt unambiguously wintry.

Atop a ridge overlooking the downtown and its seemingly endless fringe of homes, we happened across a teenage boy casually perched on the ground beside the trail and using the trunks of the leafless oaks as targets for his Christmas pellet rifle. We exchanged pleasantries, and listened to his expressions of admiration for the range of hills in which he had wandered for many years. Except for the gun, I thought, this could have been me twenty-five years earlier: I too had spent most of my youth in this same valley.

"What really bugs me," he said, "are those new houses down there," waving his outstretched hand in the direction of a recently scraped slope below us that would soon sprout with the latest California ranch styles.

Again like me, I thought. In the thirty years since my parents and their children had moved "out there" from a more central city location, I had always regretted the growth and development which had transformed much of the open grasslands, forested slopes, and orchard-covered valleys into expanses of houses, and which had changed the business district from one surrounded by empty rural landscapes into one amid and part of an urban area of a quarter of a million people. The creeks where as a youth I had searched for tadpoles or the swarms of newly transformed toads were now deep, open channels lined by high vertical walls of concrete and bordered by chain link fences. The railroad tracks along which I had walked from school in the afternoon had passed through patches of wild "middle" landscape as only an old rail line can do, but now it was gone completely, much of the length of right-of-way occupied by a new highway. The wildness of the hillslope that was formerly "the mountain" had long since been compromised by the construction of a new high school which, with a sort of poetic justice, had been subsequently closed, a victim of declining enrollments. Yes, I thought, this young Californian and I have much in common.

"So, do you live around here?" my father asked, out of politeness as much as interest. "Yeah, I live right over there," he said, turning and pointing toward another slope which might have been developed ten or fifteen years earlier. The difference between this boy and that younger me became apparent then as abruptly as had the similarity earlier. He was one of the intruders who had transformed the landscapes of my childhood, the people whose coming I regretted. We differed not in our unhappiness that a change had distorted the remembered environments of our youths, but rather in our perception of what constituted that ideal, "unadulterated" environment of the past. He, being younger and coming to the valley more recently than I, looked back with affection to a time that for me was well along toward debasement. Was there, also, a still older man or woman who could recall the time when the valley was even more remote and less developed than I could remember—a time when "my" creeks ran clear and had fish as well as toads, when the railroad was new or perhaps did not exist at all, when "the mountain" was not an unusual wooded ridge but only one of many? That person probably regretted my coming, my contribution to the development of his or her childhood landscapes. And before that person? Might there be a string of people, each remembering with fondness the images of youth and each regretting the loss of those images under the onrushing growth and development? Or, turning forward instead of backward, in the future will someone who is now a child look back upon the valley of today as ideal and condemn the development which changed the landmarks of his or her youth? Would none of these people wish for no development at all, but only for the degree of development known in childhood?

Wallace Stegner, writing in *American Places,* is unusually thoughtful about his individual perspective within the framework of time. He recognizes that the landscape that he initially inherited three decades ago in the hills south of San Francisco had been altered by two centuries of European settlement: "Changes that had taken place before our coming did not trouble us, and had not seriously damaged the hills. We accepted our surroundings, as if they had just been smoothed and rounded and peopled by the hand of God, and we tried to keep them as we found them." Stegner even appreciates that the construction of his home, and the introduction of water, garden plants, and pet animals, contributed to the "damage" even if it was not "serious." Nevertheless, he appears completely unprepared to accept the changes that the coming of subsequent settlement will bring. At first, the surroundings were lonely, with the only light at night from "stars and moon and the glow on the sky above the valley conurbia." But gradually new homes are closing in, and Stegner finds himself powerless to keep the hills "as [he] found them":

> So the pasture is likely to go, and with it will go much that has made our thirty years here a long delight, a continual course in earth-keeping. All around it the bulldozers are at work tearing up the hills for new houses—twelve here, thirty-one there, fifteen across the gully. That may be the reason why we see coyotes where we never used to see them, and why the one fox we have seen in years was in that same field. There may literally be no other place for them, or for the deer that in these winter months herd up in bunches of eighteen or twenty and come and go through our yard, as swallows hit a hole in a barn. What we see out of our bedroom windows may be only the remnants, deceptively numerous because concentrated in this one patch of open space.[64]

As Stegner first found them, the hills were neither natural nor free from development, and even then the wild nature that persisted was but a "remnant" of earlier times. Yet, like me standing on a nearby hill, and like the young boy after me, Stegner does not seem to view his appearance on the scene as a destructive disturbance. Does each individual share an image that he or she is part of a "peopling" of the environment "by the hand of God"? Do we all perceive only those who come after us as invaders, "tearing up the hills" which only we know best how to keep?

The point of presenting these experiences in this manner is not to advocate resignation when faced with the inconsistencies of the people involved. Instead, the anecdotes illustrate the dilemma inherent in assigning either positive or negative values to various landscape changes, particularly changes that involve development of wild or rural landscapes. Is the reaction to such change necessarily so personal, so determined by differences in individual lives, that no one can ever be confident of being "against development" as an uncompromising principle? Are those who express such antagonism united in general position only, and not in images of the most desirable combination of developed and undeveloped lands, or even in what types of developed landscapes are acceptable? Is there nothing about the intensive commercial development of the last three decades which makes it a special threat to the interests of those who value the rural and the wild?

In at least one sense, the disdain frequently expressed for the commercial development of the past few decades is not unique to our times. Americans have a long tradition of viewing cities as undesirable places, and of resenting the intrusion of technology into the natural or the pastoral scene. Geographers like Clarence Glacken and Yi-Fu Tuan, who are interested in the history of ideas, have in fact pointed out that anti-urban sentiments are far older than even the United States, extending back before the time of Christ in both Western and Oriental civilizations.[65] Similarly, the idealization of the farm, particularly of the family operation free from corporate ties, which often goes hand-in-hand with condemnation of the city, extends back at least as far as the Jeffersonian husbandman. The conflict between constancy and change may be part of the human condition. This conflict is an expression of the contrasting tendencies toward security and adventure, between what Daniel Luten has called the contrary personality traits of homesickness and wanderlust.[66] The character of the landscape and the human reaction to that character reflect these antagonistic tendencies which battle each other within us all.

I think back to the teenage boy I encountered in the hills on a winter day. Some of me was in that other person, the part of me that condemned the economic and population growth which had transformed my home town into a congested city. Did I deplore the change because it seemed a threat to what was familiar, to my security, to my feeling of homesickness? I suspect so. But in contrast to those negative feelings toward development, I must also admit that as a boy I held much pride in the growing town's plans for new streets, in each additional traffic light that was erected, in the big new stores, and even in the increasing prominence with which the town's name appeared on the road maps. Did I admire the change which development brought because it appealed to my appreciation for a change-of-scene, to my sense of adventure, to my wanderlust? Again, I suspect so.

These conflicting emotional needs can be satisfied only by maintaining as diverse an array of landscapes as possible—cities and wilderness, standardized commercial strips and distinctive small settlements, large corporate farms and modest family farms, suburbs and pastures, freeways and back roads. Perhaps those who criticize the trends of contemporary development have a point: It is the wilderness, the small town, the family farm, the pastures and fields, the back roads which have been losing ground. It is their futures which seem threatened.

But they are still with us today. The United States, even more than in Stewart's day, has a richness in landscapes which change has added to rather than diluted. A trip on U.S. 40 today provides the opportunity for images and experiences of 1950 as well as 1980. It is a trip still worthwhile.

NOTES AND INDEX

Notes

We used many sources for our essays. These sources included scholarly books, and articles, government documents, popular writings, our discussions with individuals, and conversations that we overheard. The list below, however, includes only those sources that are quoted in the text.

1 Carl Sauer, *Land and Life* (Berkeley: University of California Press, 1963), p. 400.
2 John T. Cunningham, *This Is New Jersey,* 3d ed. (New Brunswick, N.J.: Rutgers University Press, 1978), p. 252.
3 Writers' Program, *Maryland: A Guide to the Old Line State* (New York: Oxford University Press, 1940), p. 196.
4 Jean Holmes, "Make Her Pretty, but Don't Change Her Face," *Ellicott City Bicentennial Journal* (Ellicott City: Ellicott City Bicentennial Association, 1972), p. 4.
5 Ibid.
6 John Greenleaf Whittier, "Barbara Frietchie," *Anti-Slavery Poems: Songs of Labor and Reform* (Boston and New York: Houghton Mifflin, 1888), pp. 245–48.
7 Ibid., p. 245.
8 Writers' Program, *Maryland,* p. 282.
9 City Clerk, City of Hagerstown, letter to authors, October 22, 1980.
10 Robert C. Alberts, *Mount Washington Tavern* (Richmond, Va.: Eastern National Park and Monument Association, 1976), p. 26.
11 Robert Bruce, *The National Road* (Washington: National Highways Association, 1916), p. 55.
12 John Fraser Hart, "Land Rotation in Appalachia," *Geographic Review* 67 (1977): 148, 151.
13 J. C. Harrington, *New Light on Washington's Fort Necessity* (Richmond, Va.: Eastern National Park and Monument Association, 1957), pp. 5–6.
14 David Lowenthal, "Age and Artifact: Dilemmas of Appreciation," in *The Interpretation of Ordinary Landscapes,* ed. D. W. Meinig (New York: Oxford University Press, 1979), p. 124.
15 Alberts, *Mount Washington Tavern,* pp. 2–3.
16 Harrington, *New Light on Washington's Fort Necessity,* p. 1.
17 Lowenthal, "Age and Artifact," p. 124.
18 Elizabeth Y. Ainsworth, *Wheeling: A Pictorial History* (Norfolk, Va.: Donning, 1977), p. 14.
19 Bruce, *The National Road,* p. 81.
20 Writers' Program, *West Virginia: A Guide to the Mountain State* (New York: Oxford University Press, 1941), p. 283.
21 John A. Williams, *West Virginia: A Bicentennial History* (New York: W. W. Norton, 1976), p. 176.
22 Writers' Program, *The Ohio Guide* (New York: Oxford University Press, 1940), p. 493.
23 J. B. Jackson, *American Space* (New York: W. W. Norton, 1972), p. 63.
24 Charles B. Hunt, *Natural Regions of the United States and Canada* (San Francisco: W. H. Freeman, 1974), p. 330.
25 Display in the National Road–Zane Grey Museum, Norwich, Ohio.
26 Mary Elizabeth Wood, *French Imprint on the Heart of America* (Knightstown, Ind.: Bookmark, 1977), p. 34.
27 R. L. Baker and M. Carmony, *Indiana Place Names* (Bloomington: Indiana University Press, 1975), p. 67.
28 Joseph P. Lyford, "Vandalia," in *Prairie State,* ed. Paul Angle (Chicago: University of Chicago Press, 1968), p. 560.
29 Ibid., p. 563.
30 Paul C. Nagel, *Missouri: A Bicentennial History* (New York: W. W. Norton, 1977), p. 168.
31 Ibid., p. 172.
32 Willa Cather, *Death Comes for the Archbishop* (New York: Alfred Knopf, 1927), p. 235.
33 Wallace Stegner, *Wolf Willow* (New York: Viking Press, 1962), p. 7.
34 J. B. Jackson, "The Kansas Landscape," *Kansas Engineer* 51:2 (1967): 28.
35 Truman Capote, *In Cold Blood* (New York: Random House, 1965), p. 3.
36 Aldo Leopold, *A Sand County Almanac* (New York: Oxford University Press, 1949), p. 176.
37 Ethel M. Gillette, *Idaho Springs: Saratoga of the Rockies* (New York: Vantage Press, 1978), p. 2.
38 Ibid., pp. 187–88.
39 Writers' Program, *Colorado: A Guide to the Highest State* (New York: Hastings House, 1941), p. 277.
40 Louise C. Harrison, *Empire and the Berthoud Pass* (Denver: Big Mountain Press, 1964), pp. 451–52.
41 Marshall Sprague, *Colorado: A Bicentennial History* (New York: W. W. Norton, 1976), p. 48.
42 George M. Monahan, "Berthoud Pass—A Trail Transformed," in *Empire and the Berthoud Pass,* Louise C. Harrison (Denver: Big Mountain Press, 1964), pp. 431–32.
43 Harrison, *Empire and the Berthoud Pass,* p. 441.
44 James Kelly, "Rocky Mountain High," *Time* 116:24 (1980): 28.

45 Writers' Program, *Colorado: A Guide to the Highest State,* p. 3.
46 Ibid., p. 6.
47 Joel Garreau, *The Nine Nations of North America* (Boston: Houghton Mifflin, 1981), p. 302.
48 Writers' Program, *Utah: A Guide to the State* (New York: Hastings House, 1941), p. 8.
49 Ibid., p. 372.
50 Ibid., p. 383.
51 Christopher Springmann, "Utah's Johnny Appleseed," *National Wildlife* 13:6 (1975): 50.
52 Writers' Program, *Utah: A Guide to the State,* p. 382.
53 Ibid., p. 385.
54 Wallace Stegner, "History, Myth, and the Western Writer," in *Great Western Short Stories,* ed. J. Golden Taylor (Palo Alto, Ca.: American West, 1967), p. xiv.
55 Donald Dale Jackson, "Wilderness Will Find You," *Sports Illustrated* 41:25 (1974): 73, 76.
56 Writers' Program, *Nevada: A Guide to the Silver State* (Portland, Oreg.: Binfords and Mort, 1940), p. 125.
57 Ibid., p. 129.
58 Robert H. Ferrell, ed., *Off the Record: The Private Papers of Harry S. Truman* (New York: Harper and Row, 1980), p. 317.
59 George R. Stewart, *Donner Pass* (Menlo Park, Calif.: Lane, 1964), p. 61.
60 Ibid., p. 62.
61 Federal Writers' Project, *California: A Guide to the Golden State* (New York: Hastings House, 1939), p. 568.
62 George R. Stewart, *Earth Abides* (New York: Random House, 1949).
63 E. B. White, *Stuart Little* (New York: Harper and Row, 1945), p. 100.
64 Wallace Stegner, "Remnants," in *American Places,* Eliot Porter, Wallace Stegner, and Page Stegner (New York: E. P. Dutton, 1981), pp. 209–11.
65 C. J. Glacken, *Traces on the Rhodian Shore* (Berkeley: University of California Press, 1967); Yi-Fu Tuan, *Topophilia* (Englewood Heights, N.J.: Prentice Hall, 1974).
66 D. B. Luten, "Engines in the Wilderness," *Landscape* 15:3 (1966): 26.

Index

DESIGNED BY FRANK O. WILLIAMS
COMPOSED BY THE COMPOSING ROOM, KIMBERLY, WISCONSIN
MANUFACTURED BY MALLOY LITHOGRAPHING, INC.
ANN ARBOR, MICHIGAN
TEXT AND DISPLAY LINES ARE SET IN SABON

Library of Congress Cataloging in Publication Data
Vale, Thomas R., 1943-
U.S. 40 today.
Includes bibliographical references and index.
1. United States—Description and travel—
1981– . 2. United States—Description and travel—
1940–1960. 3. United States Highway 40. 4. Vale,
Thomas R., 1943– . 5. Vale, Geraldine R.
I. Vale, Geraldine R. II. Title. III. Title: US
forty today.
E169.04.V34 1983 917.3'04927 83-47765
ISBN 0-299-09480-4
ISBN 0-299-09484-7 (pbk.)

TODAY